More Prais

WHEN ANIMA

"Octopuses dream. Rats suffer nightmares. Chimps trained in sign language 'talk' in their sleep. This revelatory book shows us that animals' minds, like ours, are gloriously nimble, vivid, and complex, even during sleep. This is thrilling, essential reading for all of us seeking to expand our understanding of the wonder of consciousness."

—SY MONTGOMERY, *New York Times* bestselling author of *The Soul of an Octopus: A Surprising Exploration into the Wonder of Consciousness*

"*When Animals Dream* is a revolutionary book. Peña-Guzmán convincingly shows that animals as diverse as rats, monkeys, and octopuses dream, and that sometimes scientists can even tell us what they dream about. This beautiful book opens a window into the fascinating mental and emotional world-building abilities of animals, inviting us to see that we must treat animals much better than we do."

—BARBARA J. KING, author of *Animals' Best Friends: Putting Compassion to Work for Animals in Captivity and in the Wild*

"In this exciting book, Peña-Guzmán helps readers come to see that differences in animal minds do not preclude our recognizing animals as morally valuable beings. Seeing other animals as dreamers allows us to ask new questions about what it might be like to be them and to wonder about the stuff of their dreams."

—LORI GRUEN, author of *Ethics and Animals: An Introduction*

"*When Animals Dream* is a very important book that closes the door on some questions about animal minds, but more importantly opens many others for further transdisciplinary discussions about the rich inner lives and moral significance of nonhumans. Peña-Guzmán clearly shows that the question at hand isn't if animals dream, but rather why dreaming evolved as it has and what it's good for."

—MARC BEKOFF, coauthor of *The Animals' Agenda* and *A Dog's World*

"*When Animals Dream* is a fascinating, challenging, and thought-provoking book that gives human exceptionalism a philosophically-grounded middle finger."

—LEON VLIEGER, *Inquisitive Biologist*

WHEN ANIMALS DREAM

When Animals Dream

The Hidden World of Animal Consciousness

David M. Peña-Guzmán

PRINCETON UNIVERSITY PRESS
PRINCETON AND OXFORD

PUBLISHED BY PRINCETON UNIVERSITY PRESS
41 William Street, Princeton, New Jersey 08540
99 Banbury Road, Oxford OX2 6JX

press.princeton.edu

First paperback printing, 2023
Paperback ISBN 9780691227061

The Library of Congress has cataloged the cloth edition as follows:

Names: Peña-Guzmán, David M., author.
Title: When animals dream : the hidden world of animal consciousness / David M. Peña-Guzmán.
Description: Princeton : Princeton University Press, [2022] | Includes bibliographical references and index.
Identifiers: LCCN 2021050401 | ISBN 9780691220093 (hardback) | ISBN 9780691220109 (ebook)
Subjects: LCSH: Consciousness in animals. | Animal rights—Moral and ethical aspects. | BISAC: PHILOSOPHY / Ethics & Moral Philosophy | SCIENCE / Life Sciences / Zoology / Ethology (Animal Behavior)
Classification: LCC QL785.25 .P46 2022 | DDC 156/.3—dc23/eng/20211208
LC record available at https://lccn.loc.gov/2021050401

British Library Cataloging-in-Publication Data is available

Editorial: Matt Rohal
Production Editorial: Ali Parrington
Jacket/Cover and Text Design: Chris Ferrante
Production: Erin Suydam
Publicity: Matthew Taylor & Carmen Jimenez
Copyeditor: Michele Rosen

Jacket/Cover art: *Muusoctopus levis,* Enteroctopodidae. Plate LXXIX from *Die Cephalopoden* by Carl Chun, 1915.

This book has been composed in Garamond Premier Pro

CONTENTS

ACKNOWLEDGMENTS

Even if only my name appears on the cover, this book is the offspring of what the feminist science scholar Karen Barad calls an "agential network," which refers to complex structures whose effects are best understood as emerging from the convergence of multiple factors rather than the conscious intent of any one individual. In these networks, agency is decentralized and distributed such that even the most centrally located of nodes can never claim to be more than that—a node, one among many.

I want to express my gratitude to the many nodes that have made this book possible, beginning with two people who, without their knowledge, put me on the path that would culminate in me writing these words. The first is Tanya Augsburg, whose invitation to speak at the 2018 meeting of The Animal Union marked my first public mention of my interest in the nightly experiences of other species—even if, at the time, this interest was little more than a nebulous idea in the back of my mind. The second is Marjolein Oele, who invited me to give a talk at the University of San Francisco a few weeks later, in April 2018. I used this opportunity to dig more seriously into the science and philosophy of dreaming and sculpt my still ill-defined interest into something resembling a coherent philosophical thesis. This talk was well received by students and faculty alike, which is how I started toying with the idea of writing a book.

Having never done such a thing, however, the thought filled me with dread and made me ooze more insecurities than I care to admit—about my writing style, about my authorial voice, about my research skills, and, of course, about being found out as the impostor that I obviously was. Fear grabbed hold of me, and I decided to simply let the project fall by the wayside.

It was Rabih Hage who changed my mind and convinced me not to flee from the challenge. It was him who, with the skill of a seasoned therapist, assuaged my fears and encouraged me to write, even when I got my first real taste of writer's block. It was also him who, with the generosity of a partner but the rigor of an expert, answered all my questions about neuroscience while pressing me with tough questions of his own about the philosophical use I intended to make of it. Sadly, this kindness backfired on him since it was him, too, who endured more rants about animals and their dreams in a single year than anyone should in an entire lifetime—a suffering he bore with the patience of a saint. Throughout this process, he has been it all: my lover, my friend, my interlocutor, my editor, my confidant, my critic. This book has been made better by him, as have I. I dedicate this book to him, my partner in all things.

Writing can be an unbearably lonely activity, and most of the writing for this book was done under conditions of intense isolation: during confinement in Paris, France, in 2020. These were difficult times that I braved by holding on for dear life to my partner, family, and friends. My daily interactions with my partner anchored and sustained me. My phone calls with my mother, my brother, and my extended family in Mexico drew me out of myself and gave me perspective. My friendships renewed and restored me.

Many of these friendships directly aided and abetted in the writing of this book. Jessica Locke, Osman Nemli, and I formed a writing accountability group that met weekly during the pandemic. I benefitted tremendously from these encounters, which gave me structure, kept me on track, and kept me honest. I thank both for their constructive and critical feedback on various chapters. I also thank Rebecca Longtin, Joel M. Reynolds, Alex Feldman, Michael Sano, and Deborah Goldgaber, all of whom also lent their support. Their observations, critiques, and recommendations had a meaningful impact on my thinking and writing. Special thanks to Rebekah F. Spera, who put together the book's index and edited the manuscript from top to bottom, saving readers from some of my less honorable writing habits along the way.

I also would like to thank the members of two scholarly communities at San Francisco State University that helped me process my ideas in a welcoming and collaborative environment: the "Historicity of Consciousness" reading group that I co-founded with Arezoo Islami, and the STS HUB run by Laura Mamo, Martha Kenney, and Martha Lincoln. Also deserving of mention are my colleagues in the School of Humanities and Liberal Studies: Cristina Ruotolo, Tanya Augsburg, Jose Acacio de Barros, Denise Battista, Sean Connelly, Karen Coopman, Brad Erickson, Mariana Ferreira, Judith Fraschella, Laura Garcia-Moreno, Logan Hennessy, George Leonard, Sarah Marinelli, Marie McNaughton, Peter Richardson, Steve Savage, Mary Scott, Nick Sousanis, Christopher Sterba, Shawn Taylor, Rob Thomas, and Stacey Zupan. I could not have finished this book without the material support of the George and Judy Marcus Fund for Excellence in the Liberal Arts, which financed my sabbatical leave in the spring of 2020.

Finally, I take my hat off to the very competent team at Princeton University Press. Matt Rohal has been an efficient, wonderful, and caring editor who believed in this project even when I still harbored serious doubts about its viability. He saw its potential and nudged me to make it accessible to a general audience, something that doesn't come naturally to people in my line of work (academic philosophy). Michele Rosen proved an excellent copyeditor whose eagle eye for detail improved the manuscript greatly. Ali Parrington saw the manuscript through copyediting and the successive production stages, ensuring that all the deadlines were met by all the relevant nodes. Chris Ferrante designed the stunning cover, while Emma Burns took charge of the in-chapter illustrations under the coordination of Dimitri Karetnikov.

Their artistic talent has added an entire dimension of meaning to the book for which I take no credit.

Like me, each of these individuals is a node in the network that begat the book you now hold in your hands. Still, any errors unearthed in the book are through no one's fault but my own.

INTRODUCTION

In the Trenches of Sleep

I can hear little clicks inside my dream.
Night drips its silver tap
down the back. At 4 A.M. I wake. Thinking
—ANNE CARSON[1]

HEIDI'S DREAM

Season thirty-eight, episode one of the PBS series *Nature*, "Octopus: Making Contact,"[2] promised viewers a rare journey into the inner lives of octopuses, billed as "the closest we may get to meeting an alien." The star of the one-hour documentary is Heidi, a female day octopus (*Octopus cyanea*) who lives with the narrator, David Scheel, a biologist at Alaska Pacific University. Unlike most captive octopuses, Heidi lives neither in an aquarium nor in a laboratory, but in Scheel's private residence in Anchorage—a charming mix of roommate, companion animal, and research assistant.

"Octopus: Making Contact" tells a tale of octopuses not as "stupid creatures," which is how the Greek philosopher Aristotle described them in 355 BCE, but as intelligent and naturally curious beings who have unique personalities, recognize others of the same species, and solve complex problems. From start to finish, octopuses are presented as conscious agents who know when they are being observed and who, more importantly, do not hesitate to observe in return. "When you look at them," says Scheel, "you feel like they're looking back. That's not an illusion. They *are* looking back."

Near the end of the documentary, as Heidi is shown sleeping in her tank, Scheel reports: "Last night, I witnessed something I've never seen recorded before." What follows is a breathtaking one-minute-long shot. In it, Heidi is at first peacefully restful, but after a few seconds her skin lights up, displaying a sequence of dramatic, multicolored patterns, each one more mesmerizing than the last. The "something" Scheel is referring to may be an *octopus dream*.

His voice then walks the viewer through each of Heidi's arresting displays, noting, "you could almost just narrate the body changes and narrate the dream."

DISPLAY 1

Heidi changes from a smooth and consistent alabaster white to a flashing yellow with blotches of mandarin orange. "So here she's asleep, she sees a crab, and her color starts to change a little bit."

DISPLAY 2

From these splendid shades of yellow and orange, Heidi changes to a dark and piercing purple, a purple so deep that for a fraction of a second, we cannot tell where her body ends and the dark blue background begins. "Octopuses will do that when they leave the bottom," usually after a successful kill, Scheel explains.

DISPLAY 3

Heidi then changes into a series of light grays and yellows, except this time the colors are crisscrossed by a disordered topology of ridges and spiky horns, the textured byproduct of the contractions of the papillae on her skin. "This is a camouflage, like she's just subdued a crab and she's just going to sit there and eat it, and she doesn't want anyone to notice her."[3]

The camera then turns to Scheel himself, who says with noticeable elation: "This really is fascinating [. . .] If she's dreaming, *that's* the dream."

Heidi became a viral sensation overnight. Within days, thousands of people shared the video of her dream on social media, and major news outlets rushed to cover the story. Viewers were simultaneously fascinated and stupefied. Her sleep displays were stunning, a veritable kaleidoscope of flesh. But what did they mean? And beneath this procession of color and texture, what was Heidi herself thinking or feeling? As Elizabeth Preston put it in the *New York Times*, "[A]n octopus is almost nothing like a person. So how much can anyone really say with accuracy about what Heidi was doing?"

Pan out and the bigger question becomes: What goes on in the minds of nonhuman animals when they sleep, or, as

FIGURE 1. Heidi displays three separate chromatic patterns in a row while asleep, probably on account of experiencing a dream in which she is hunting and eating prey.

the poet Anne Carson says, when "night drips its silver tap"? Do they experience those penetrating nightly visions that humans do, which Shakespeare described as "the children of an idle brain"? Or do their minds simply plummet into a psychic void in which no conscious experience takes root? Can other animals—not just octopuses, but parrots, lizards, elephants, owls, zebras, fish, marmosets, dogs, and so on—truly dream? If so, what does this tell us about who these creatures are and how they dwell in this world? And if not, does this mean that dreaming may be the cognitive Rubicon that separates us from the other animals? Are humans "the dreaming animal," as the Spanish philosopher George Santayana believed?[4]

This book is about these questions.

ANIMAL INTERIORITY

Even though humans have been fascinated by the possible dreamworlds of other animals for millennia,[5] one of the first modern scientific publications devoted to animal dreaming appeared in 2020. In an article published in the *Journal of Comparative Neurology* under the title "Do All Mammals Dream?," the biologists Paul Manger and Jerome Siegel express doubt that only humans experience dream sequences during sleep, and they wonder whether dreaming—that curious mental happening that the sociologist Eugene Halton describes as "the mind's nightly ritual of inner icons"[6]—may be a universal feature of mammalian life, something we share with all other species whose young feed from the mother's mammary glands. I will come back to this mammalocentric hypothesis in chapter 1, but for now I want to emphasize that this article stands out within the field of animal sleep research as a genuine anomaly: a publication in a scientific journal that uses the terms "dream" and "dreaming" explicitly in connection to animals other than *Homo sapiens*.[7]

To be clear, this is not the only publication to shed light on what goes on inside the minds and bodies of animals during sleep. Far from it. Over the last century, biologists, psychologists, and neuroscientists have made significant strides in cracking the code of animal sleep, giving us a fuller picture of the imperatives of animal experience across the great sleep-wake divide. Nevertheless, these same experts have historically shied away from describing their findings using the language of dreams. Instead, they have opted for more phenomenologically ambivalent terms, such as "oneiric behavior"[8] and "mental replay,"[9] that allow them to talk at great length about the mechanics of animal sleep—the biological processes that regulate it, the physiological changes that prompt it, the neurochemical changes it occasions, and so on—without needing to take a stance on whether any of the animals under study actually experience anything subjectively at any point during the cycle of sleep. Because of their intrinsic agnosticism, these terms end up blotting out some of the most philosophically stimulating questions raised by the possibility of animal dreaming, especially questions concerning consciousness, intentionality, and subjectivity.

In this book, I build on contemporary animal sleep research to show that what scientists refer to as "oneiric behaviors" and "mental replay" in sleeping animals should be interpreted as the result of internally generated dream sequences that animals experience—even if only momentarily—as their very reality. Rejecting this phenomenological interpretation, I argue, would require holding two conflicting beliefs at once: first, that many animals display the same patterns of motor and neural activity during sleep that are widely accepted as indices of dreaming in humans; and second, that while this bustle is going on inside them, these same animals sense, feel, and think nothing. It would almost require believing that the minds of animals magically disappear into the ether the moment animals drift off into sleep; that, immediately upon entering the kingdom of Hypnos, a gaping abyss opens up beneath them and swallows them whole. While this position is not necessarily illogical, a close reading of the empirical data reveals it to be untenable. Even if scientists are reluctant to talk about the dreams of animals (say, for reasons of scientific humility), their findings point in precisely that direction.

My concern is that, aside from betraying a problematic double standard,[10] this reluctance to talk about animal dreaming feeds a larger cultural prejudice that rationalizes our appalling treatment of animals. In a seminal article on animal consciousness, the father of cognitive ethology, Donald Griffin, called this prejudice "mentophobia"—the fear of viewing animals as creatures with minds of their own.[11] This fear leads us to see animals as food to be consumed, reservoirs of labor power to be exploited, resources to be used, and specimens to be cultured and dissected—as anything, that is, except creatures who live, feel, and think on their own terms. While mentophobia affects all areas of social life, Griffin recognized that it exerts an exceptionally strong pressure on the scientific community, a pressure that is most conspicuously on display whenever scientists resist attributing complex mental states to the animals they study even when there is ample support for it. It is because of mentophobia that most of us continue to see animals, in the now infamous words of the philosopher Normal Malcolm, as "thoughtless brutes"; that is, as creatures who eat, sleep, and die, but who never develop a meaningful cognitive, emotional, or existential bond with the world.[12] Once animals are pigeonholed into this category, their fate is sealed. There are simply too many things one cannot expect from a thoughtless brute.

One of them is the capacity to dream.[13]

And yet: watching the displays of Alaska's most famous cephalopod feels very much like witnessing the collision of two subjective realities—one human, one not. It is almost as if Heidi's flamboyant metamorphoses bring within the reach of our human, all-too-human senses that alluring yet inscrutable realm of reality from which every human observer has been barred from time immemorial: the inner world of another animal. Perhaps a phenomenology of animal dreaming can explain why. If, while watching Heidi's displays, we feel that we are coming face-to-face with another subjective reality that is recognizable and alien at once, this may be because the band of colors marching rhythmically on the surface of her skin bespeaks a dream, a dream that—much like the dreams of the myriad other animals we will encounter throughout this book—is itself an irrefutable sign that *there exist, alongside ours, endless other worlds—utterly "Other," inhuman worlds. Enigmatic, foreign, hidden animal worlds.*

Worlds without human contours.

Worlds with nonhuman centers.

AN INTEGRATIVE APPROACH

There are experts who worry that attributing dreams to animals anthropomorphizes them by projecting a uniquely human ability onto them. In their view, animal researchers should stick to what the philosopher of science Peter Winch calls "external descriptions" of behavior, leaving considerations of animal interiority to their colleagues from across the quad: the philosophers.[14] In defense of this division of intellectual labor, they offer a host of arguments. Sometimes, they invoke the authority of "Morgan's canon," which says we must opt for the simplest possible explanation of animal behavior.[15] Sometimes, they appeal to the philosophical "problem of other minds," which maintains that we cannot say that animals have an interior life because we lack direct access to their first-person experience of the world.[16] At other times, however, they hint at the problem of language. In the absence of a shared language, they say, we cannot make empirically meaningful claims about how, when, or why—or even whether—other animals dream, let alone about the nature, structure, and quality of their putative dream experiences. What are dreams, after all, if not unobservable mental happenings whose existence we can infer only on the basis of subjective verbal reports—reports that animals cannot provide?

However appealing, this view relies on the conceit that the scientific study of dreams depends solely or mostly on the compilation, analysis, and interpretation of dream reports. Surely, dream scientists have learned, and continue to learn, a great deal from the verbal reports of human dreamers about what our minds and bodies do when we go "offline." But the bulk of dream research since the 1980s has not been exclusively (or even primarily) based on the analysis of linguistic reports. It has been based on the investigation of the neural and behavioral correlates of dream experiences, which is to the say, the brain activity and bodily behaviors that correspond with the subjective experience of dreaming. A brief survey of contemporary human dream research reveals a vast, interdisciplinary, and rapidly evolving field in which experts concentrate on spotting the neural signatures (e.g., ponto-geniculo-occipital, or "PGO," waves)[17] and behavioral markers (e.g., rapid eye movements or "REMs") of human dream phenomenology.[18]

While our inability to speak with other animals certainly limits what we can know about their dream experiences, it does not prevent us from making meaningful and empirically

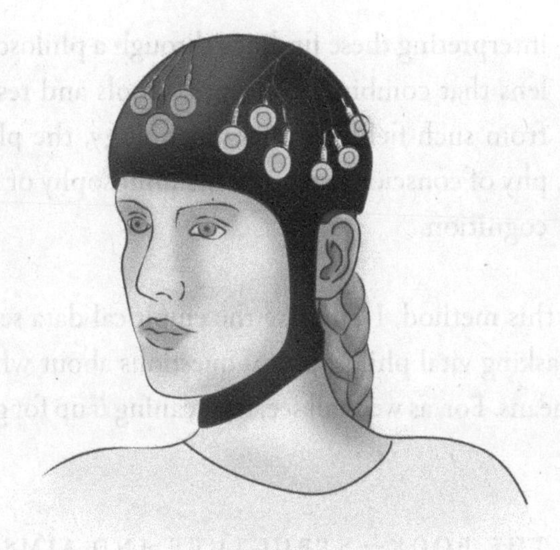

FIGURE 2. While linguistic reports remain a valuable tool in dream science, much contemporary dream research relies on the use of electroencephalography (EEG), functional magnetic resonance imaging (fMRI), and positron emission tomography (PET) to isolate the neural circuits involved in dreaming. Here, a woman wears an EEG headset in preparation for a study.

educated claims about their capacity to dream, or even from ruminating about the possible implications of this capacity for ongoing scholarly debates about animal consciousness, animal emotion, and animal ethics.[19] Indeed, throughout this book I use an *integrative method* to advance several such claims. In essence, this method involves:

1. surveying the empirical literature on animal sleep for findings that might point to dream experiences in other animals; and,

2. interpreting these findings through a philosophical lens that combines conceptual tools and resources from such fields as phenomenology, the philosophy of consciousness, and the philosophy of animal cognition.

Using this method, I can take the empirical data seriously while asking vital philosophical questions about what this data means. For, as we shall see, its meaning *is* up for grabs.[20]

THE BOOK—STRUCTURE AND AIMS

People who interact with animals as part of their everyday life—animal lovers, farmers, veterinarians, animal activists, and so on—may be tickled by the thought that someone would write an entire book about something that may strike them as obvious: that we share the ability to dream with many other critters. But holding this belief is one thing; defending it on scientific grounds is another; and teasing apart its philosophical implications is yet another. In the chapters that follow, I do all three.[21]

In chapter 1, "The Science of Animal Dreams," I turn to animal sleep research to catalog evidence that animals run "reality simulations" during key phases of their sleep cycles. Even taking certain methodological and conceptual limitations into account, the preponderance of this evidence supports the conclusion that humans are not the only dreamers on earth.

In chapter 2, "Animal Dreams and Consciousness," I consider the philosophical significance of the evidence laid out in chapter 1. Here, I introduce the "SAM" model of consciousness, which distinguishes three types of awareness: "S" for subjective (being at the center of a phenomenal field of experience), "A" for affective (experiencing events as emotionally shaded), and "M" for metaconscious (having the ability to reflect upon one's own mental life). Guided by phenomenological theories of dreaming, I assert that all animals who dream are necessarily *subjectively conscious*, that most (if not all) are also *affectively conscious*, and that a select few may be *metaconscious* as well.

In chapter 3, "A Zoology of the Imagination," I take the discussion of animal consciousness to a higher level by accentuating the imaginative character of dreams. Given that dreams hinge on the generation of sensory (visual, tactile, auditory, and so on) imagery, creatures who dream must possess what the philosopher of mind Jonathan Ichikawa calls "imaginative capacities," such as creativity, fantasy, and make-believe. I explore how these capacities congeal in dreams while presenting dreams as part of a larger spectrum of imagination that includes, inter alia, hallucinations, daydreams, and mind-wanderings.

In chapter 4, "The Value of Animal Consciousness," I tackle the ethical dimension. Do the dreams of animals matter from an ethical standpoint? Under most ethical frameworks, the answer to that would be yes, as consciousness is thought to determine which entities have moral status and which do not. Here, I use the philosopher Ned Block's famous theory of consciousness as a jumping-off point for articulating a novel account of why dreams are pregnant with what I call "moral force." On this account, dreams are morally significant because they reveal animals to be both carriers and sources of moral value, which is to say, beings who matter and *for whom* things matter.

The book closes with a short epilogue, "Animal Subjects, World Builders," in which I offer some final thoughts about the subjectivity of other animals and about what binds us to and cleaves us from them. It is in this tension between sameness and difference, between conjunction and disjunction, that the heart of this book lies. If inhabited correctly, I argue, this tension can open up contemporary debates about animal minds and animal experience and make us question some of our more disturbing assumptions about our nonhuman comrades, so that we can begin the task of collectively learning to see animals truly anew—no longer as the evolutionarily, cognitively, metaphysically, or even spiritually impoverished versions of us that we have historically taken them to be, but as the fully realized, inviolable, sacred versions of themselves that they already are and always have been.

CHAPTER 1

The Science of Animal Dreams

As dogs, cats, horses, and probably all the higher animals, even birds, have vivid dreams [. . .], we must admit that they possess some power of imagination.
—CHARLES DARWIN[1]

"THE SILENT CENTURY"

Scientific debates about animal dreams date back at least to the late nineteenth century. In the wake of the publication of Charles Darwin's *On the Origin of Species* and *The Descent of Man*, proponents of the rising evolutionary framework began circulating the idea that animals share many of the mental capacities previously assumed to be solely human, including the capacity to dream.

One of the earliest proponents of this idea was the Scottish physician William Lauder Lindsay, who offered a spirited defense of it in his 1879 book *Mind in the Lower Animals in Health and Disease*. Citing classical and contemporary reports of dreaming animals, Lindsay held that dreaming is not the sole province of *Homo sapiens*. In a chapter entitled "Dreams and Delusions," he has the following to say about what happens in the minds of dogs when they fall asleep:

> As regards the *dog*, and especially sporting dogs, such as the harrier, the following facts have been noted, or the following inferences drawn. It appears to hunt in its dreams, as was long ago remarked by Seneca and Lucretius. During sleep movements of the tail and paws, sniffing, growling, barking, occur. There is every reason to believe that there is frequently during sleep in the sporting dog an *imaginary* pursuit of imaginary game; that this supposed pursuit gives rise to actual physical and mental excitement, including, for instance, eagerness and panting for breath caused immediately thereby; and that this excitement sometimes causes the animal to awake.[2]

And this is not limited to one or two breeds. He continues:

> Just as hounds of harriers chase in their dreams imaginary game, collies or other dogs worry in their sleep imaginary enemies or snap presumably at imaginary flies or other insect tormentors. In other words, in their sleep or dreams they appear to engage in imaginary quarrels, games, pursuits, and attacks.[3]

The chapter goes on to describe dream states in animals as varied as horses, birds, and cats and offers an elegant analysis of the relationship between dreams, delusions, and hallucinations. Dreams, for Lindsay, are powerful indications that animals have complex minds.

Belief in animal dreams was widespread at the height of the Victorian era. The antivivisection movement was gaining steam in Europe and North America, and public attitudes about the status of animals were changing rapidly.[4] In this climate, the conditions were ripe for increased interest in the mental and emotional lives of animals. Among the scientists of the time, this interest expressed itself as a general openness to a wide variety of claims—some more empirically grounded than others—about animal experience, including claims about what happens to animals when they sleep. This belief was so widespread that Darwin's protégé, the evolutionary biologist George Romanes, cited Lindsay's theory of animal dreams enthusiastically in his 1883 masterpiece *Mental Evolution in Animals*.

In this book, which was read with gusto by audiences on both sides of the Atlantic, Romanes went further than Lindsay in asserting that dreaming proves that animals are endowed with the faculty that the German moralist Immanuel Kant had categorically denied them only a hundred years earlier: the faculty of imagination.[5] Dreaming proves that animals have what Romanes dubs "imagination in the third degree,"[6] which enables animals to form mental images "independently of any obvious suggestions from without."[7] In Romanes's view, the same mental operations are needed to dream of something as to visually imagine it since, in both cases, the mind directs itself toward something absent and relates to it *as if* it were present. Dreaming, he concludes, "constitutes certain proof of imagination belonging to [. . .] the third degree."[8] In holding this view, the biologist from Kingston, Ontario, was not challenging the spirit of his time—he was channeling it.

In 1888, only five years after the publication of *Mental Evolution in Animals,* the popular magazine *The Century* ran an article about the science of dreams, nightmares, and somnambulism that included a section on animal dreams. Among the experts mentioned as defenders of an interspecies theory of dreams were lesser-known figures such as William Lindsay and Georges Romanes, as well as such luminaries as Charles Darwin.[9] A year later, the Canadian biologist Wesley Mills published his seminal *Textbook of Animal Physiology,* which included a lengthy discussion of the dreams of animals, especially dogs. That same year, the French psychologist Alfred Binet, inventor of IQ testing, reviewed several books on dreaming in the journal *The Psychological Year* [*L'Année Psychologique*], including the Italian psychologist Sante De Sanctis's book *Dreams: Psychological and Clinical Studies* [*I sogni: Studi Psicologici e Clinici*], which devoted an entire chapter to interviews De Sanctis conducted with breeders, farmers, hunters, and circus trainers about the dreams of "superior animals" such as dogs, horses, and birds.[10]

Animal dreams may have been deeply ingrained in the cultural and scientific imaginaries of the nineteenth century, but the tide eventually turned. Due to several developments, especially the rise of behaviorist psychology, what began in the 1870s as a wave of support for the complexity of animals' minds morphed over the span of only a few decades into a pervasive skepticism about animal cognition of any kind.[11] After the turn of the century, the life sciences adopted a new attitude—a colder, more distant attitude—that led new generations of scientists to distance themselves from their predecessors and to accuse them of projecting human abilities onto animals.[12] By the 1930s, many of the topics that had galvanized nineteenth century naturalists—animal reasoning, animal language, animal emotions, animal play, and, of course, animal dreaming—had fallen into scientific ill repute, and most of them remained there for a long time. I call the period stretching from the 1900s to the 1980s "the silent century" because, during this time, discussions of animal consciousness came to a standstill from which our scientific culture is still trying to break free.

Thankfully, scientists from various fields have started reclaiming some of these topics as legitimate objects of scientific inquiry. Since the 1990s, research on animal emotions has exploded, as has empirical and philosophical work on animal cognition. Unfortunately, the topic of animal dreaming has not been so lucky. As of the time of this book's publication, almost a hundred and fifty years after the publication of Lindsay's *Mind in Lower Animals*, the bulk of the scientific community continues to dismiss the idea that animals dream (let alone that their dreams might be empirically studied) as anthropomorphic, which is to say, as a romantic and nonscientific illusion that misleads us into projecting the traits of the human onto the nonhuman. This is true of many scientists who specialize in dreams, but it is almost universal among experts in animal sleep.[13]

The irony is that over the last three decades, the life sciences have generated a good deal of evidence that our nineteenth century forebearers may have been right about what the minds of animals do, in the words of Jennifer Dumpert, "at the edges of sleep."[14] In this chapter, I catalog and analyze this evidence, dividing it into three categories: electrophysiological, behavioral, and neuroanatomical. When properly interpreted, this evidence shows that our collective error was not that we considered humans and the other animals on a continuum of mental activity during the nineteenth century, but rather that we turned our backs on this continuist perspective in the twentieth and, as a result, our perception of animals changed for the worse. We began seeing their lives as so deficient, so dull, so bare, and so contemptible in comparison to ours that, in an act of collective self-delusion, we convinced ourselves that they could not possibly have what we have: a meaningful inner world. *That* was our mistake.

ELECTROPHYSIOLOGICAL EVIDENCE: FROM ZEBRA FINCHES TO ZEBRAFISH

A Soundless Song

In the year 2000, the biologists Amish Dave and Daniel Margoliash published a report in the journal *Science* describing their research on zebra finches (*Taeniopygia guttata*), which are passerine birds native to Australia. One of the evolutionary challenges these birds face is that they must learn their song from their parents and siblings, since it is not innate.[15] Research on birdsong has historically focused on what these animals do while awake to imitate and memorize their song, but Dave and Margoliash wondered whether sleep might also play a role in song acquisition. Could sleep help juvenile finches internalize the acoustic patterns they hear from their family members and commit them to long-term memory? Could these birds learn their song at least in part by rehearsing it in their minds while asleep?

To test this possibility, Dave and Margoliash performed an experiment in which they mapped the patterns of neural activation elicited in the "birdsong system" (the forebrain nucleus robustus archistriatalis) of a group of juvenile finches while they slept. By analyzing these patterns, they discovered that the brains of zebra finches oscillate between two states during sleep: a state of constant but low-level neural activity in which nothing remarkable happens, and a state marked by spontaneous bursts of high-level neural activity recurring at regular intervals. In itself, this discovery was not particularly groundbreaking, as it merely confirmed previously published work on avian sleep reporting a sleep cycle divided into phases of low and high neural activity (just like mammalian sleep). However, Dave and Margoliash then decided to map the neural pattern that emerged in the birdsong system when the finches practiced their song while awake and to compare it to the patterns elicited during sleep. Their findings were astonishing.

They discovered that the pattern elicited by the act of singing while awake was an exact structural replica of the pattern elicited during the period of sleep marked by sudden bursts of high-level neural activity, which was all they needed to realize that the brains of the zebra finches were doing the exact same thing—their neurons were firing in the same organized manner—when the birds sang their song in the middle of the day for the world to hear as when they entered a period of high neural activation during sleep. The match was so perfect that the authors realized they could map these patterns onto one another syllable-by-syllable, nay, *note-by-note*. From this, they concluded that zebra finches learn their song not only by practicing it out loud while awake ("play") but also by mentally replaying it while asleep without making a chirp ("replay"). "Replay," they write, "generates coherent activity throughout the song system that is similar to singing in the absence of actual sound production and perception."[16]

One might argue that the activation of the birdsong system during sleep is evidence that zebra finches were dreaming of singing their song. Curiously, Dave and Margoliash resist this interpretation. Instead, they argue that the replay they observed in the finches represents nothing more than the execution of a computational process (in their words, an "algorithmic implementation") that unfolds without the finches' conscious awareness.[17] In their view, finches do not experience replay any more than my laptop experiences running Adobe

FIGURE 3. The pattern of brain activity that zebra finches display when they sing their song in the waking state matches the pattern they display when they rehearse their song in the sleep state. The match is so perfect that scientists can map the patterns onto each other note-by-note.

Reader or Microsoft Word, since replay itself is a brain state wholly devoid of what Ned Block calls "experiential properties." Bluntly put, it has no accompanying phenomenology.[18]

In my view, this algorithmic interpretation is not entailed by the data so much as superimposed upon it—and, I should add, without much in the way of justification. It is not at all clear why Dave and Margoliash interpret replay algorithmically, especially when their own findings suggest that it represents a lived reality that zebra finches experience from a first-person perspective during sleep.

Two pieces of evidence, I argue, tip the balance against computationalism in favor of phenomenology. One is *temporality*. Aside from finding structural parallels between play and replay, Dave and Margoliash also found temporal ones. It took the finches roughly the same amount of time to sing their song while awake as it did to rehearse it during sleep. This is important since there is no *prima facie* reason why a computational process would need to run on the same timescale as the subjective experience it mechanically reiterates. This temporal parallelism could be the result of a shared underlying phenomenology linked to the animals' experience of lived time. If it takes these animals the same amount to time to play their song while awake as to replay it in their sleep, this could be because play and replay instantiate a similar subjective experience.[19]

The other piece of evidence here is *embodiment*. Dave and Margoliash noticed that it was not only the brain that was conscripted into the performance of replay, but the body as well—especially the throat. During replay, the birds' vocal cords expanded and contracted exactly as they did during play, which can only mean one thing: as the birds went through the steps of mentally rehearsing their song during sleep, they also practiced the bodily skills needed to actualize this song. Granted, the movements of the vocal cords did not yield any sound, but the fact that they occurred at all indicates that the memories the animals were retrieving during replay were inexorably embodied. During replay, the animals were not remembering *that*; they were remembering *how*. And it is highly likely that in the process of remembering how, the birds had a genuine auditory experience since the auditory regions of their brains also lit up like a Christmas tree. The sleeping finches, it seems, "heard" their own song in the heavy silence of sleep.

Thus, I agree with Dave and Margoliash's observation that replay takes place in the absence of "sound production," but not with their assertion that it also occurs in the absence of "perception." There is a massive difference between the two. Sound production is an objective state of affairs: did the animals sing? Did they produce sound waves? Perception is about a subjective state: did they hear a song? Did they experience sounds? My reading of the data is that the birds did not sing for the simple reason that they emitted no sound, but they *did* hear a song—what I am calling a "soundless song." They heard it silently, much like we hear the clamoring soundscapes of our own dreams—the voice of a lover, the rustling in the trees, the sound of a church bell in the distance. Unfortunately, Dave and Margoliash cannot see this because they are committed to a computationalist interpretation of replay, which I believe causes them to miss the phenomenological significance of their own findings. Zebra finches memorize their song not merely by sleeping on it, but by *dreaming* about it as well. As the seventeenth century poet John Dryden wrote in his 1665 play, *The Indian Emperor*: "The little birds in dreams their songs repeat."

Spatial Dreams

Although Dave and Margoliash's comments about the absence of perception during bird sleep can prompt skepticism about animal dreaming, we find a refreshing alternative to their algorithmic interpretation of replay in a study conducted in 2001 by Kenway Louie and Matthew Wilson from the Massachusetts Institute of Technology (MIT). While working at MIT's Center for Learning and Memory, Louie and Wilson sought to gain a better understanding of how sleep affects memory and spatial reasoning in rats. To do so, they decided to study how rats would mentally deal with a spatial task while awake and while asleep.

They selected a spatial task because rats, like humans, have a sophisticated space-mapping system in the hippocampus made up of CA1 pyramidal cells that map the animal's physical environment and fire differentially depending on the animal's position in space.[20] When a rat occupies position X in the mapped environment, a specific set of CA1 cells fire; but as soon as the rat moves to position Y, a different set of CA1 cells begin firing. Critically, if the rat then returns to position X, the exact same set of CA1 cells that originally fired will fire again. As long as the rat's physical environment remains relatively constant, researchers can pinpoint the rat's physical location with extreme precision based solely on hippocampal activation information. By tracking hippocampal activity, then, Louie and Wilson could track the physical location of the rats while they were awake, as well as the location the rats *thought* they occupied while asleep.[21]

The experiment began by acclimating a group of rats to an elevated circular track and training them to run "from a start location to a goal location for a food reward."[22] Once the rats learned this route, Louie and Wilson tracked single-cell activity in the hippocampus and recorded the specific pattern of CA1 pyramidal cell activation that resulted from their movement, thereby mapping "the sequence in which the animal's behavior [took] it through the task environment."[23] They called this pattern "RUN" since it was produced by the act of running towards the reward. Then, they wondered whether the pattern of pyramidal activation associated with RUN might re-appear during REM sleep, so they let the rats take a nap after running the track and recorded hippocampal activity while they slept. They called this second pattern "REM" since it occurred while the rats were in REM sleep. Thus, in this context RUN and REM refer to neural patterns: one connected to physically running a track while awake, and the other connected to mentally replaying this act during REM sleep.

So, what did Louie and Wilson learn? Echoing Dave and Margoliash's findings about birdsong, they discovered that RUN and REM are mirror images of one another, meaning that when the rats fell asleep, they probably dreamed about the spatial test they had just completed. Furthermore, Louie and Wilson found that RUN and REM unfolded "at approximately the same speed."[24] Just as zebra finches replayed their song while asleep in the same amount of time that it took them to play it while awake so, too, rats performed RUN and REM on a comparable timescale, on the order of minutes to seconds. Structurally and temporally, "REM *recapitulates* RUN."[25]

This is where things get interesting. Technically, Louie and Wilson merely replicated Dave and Margoliash's findings about the structural and temporal parallels that connect waking and sleeping states, but their interpretation of these parallels could not be more different. In contrast to Dave and Margoliash, who interpreted replay in zebra finches as an unconscious algorithmic process without experiential properties, Louie and Wilson interpret replay in rats as a phenomenologically rich experience—in other words, *as a dream*. Replay must be a lived reality for the rats, they say, because it depends on a past experience and because it mirrors this experience structurally and temporally.[26] Instead of a phenomenologically hollow state, replay is a genuine subjective experience "despite the absence in REM of the explicit sensorimotor cues that drive distinct neural patterns during RUN."[27] Even without any of the sensorimotor cues that they had while running the track in the waking state (such as visual information about the environment, the feeling of the ground beneath their feet, and the smell of the food reward at the end of the track), the sleeping rats *experienced* running toward the reward. They generated an internal simulation that "re-activated" or "re-constructed" a waking behavior.[28] And, of course, to reconstruct a waking behavior during sleep simply *is* to dream about that behavior, so this amounts to saying that the rats were dreaming of running the track.[29]

Let's be clear about this: the dispute between Louie and Wilson on the one hand and Dave and Margoliash on the other is not scientific. It is philosophical. At stake in it is the question of the kinds of beings that rats *are*. Are they furry little computers that implement algorithms? Or are they conscious subjects with an inner phenomenology, subjects who perceive, feel, and think? This is not a question that can be answered on purely empirical grounds, which is why these researchers can agree about the facts yet disagree about what these facts ultimately mean. The crux of their disagreement is reflected in the terms they use to describe what they see: "algorithmic implementation" versus "internal simulation." The first places replay squarely within in the ambit of computationalist theories of mind, whereas the second portrays it less as a program that the rat brain runs unconsciously and more as an experience that rats live through while asleep with the full thrust of their being.[30]

From Air and Earth to Water: Finny Dreams

More recently, in 2019, an international group of scientists from the United States, France, and Japan led by Louis C. Leung from Stanford University published an article in *Nature* titled "Neural Signatures of Sleep in Zebrafish." Like all teleost fish, zebrafish lack a neocortex, which makes the study of the neural correlates of their behaviors somewhat tricky. But they have a dorsal pallium, which ichthyologists now consider the functional equivalent of the new mammalian cortex.

By studying activation of the dorsal pallium under various conditions, Leung and his team discovered that zebrafish experience two states of sleep. One state is "slow bursting sleep" (SBS), which shares important physiological characteristics with the slow-wave sleep of mammals, birds, and reptiles, such as low frequency but synchronous neural activity and reduced eye, cardiac, and respiratory activity. The other is "propagating wave sleep" (PWS), a state reminiscent of mammalian REM sleep.[31] During PWS there is high frequency but asynchronous activity in the dorsal pallium, as well as increased but irregular cardiac activity. This state is also characterized by ponto-midbrain-telencephalic (PMT) waves, which the authors argue are the fish version of the ponto-geniculo-occipital (PGO) waves that mark the onset of REM sleep in mammals, humans included. To their own shock, the authors discovered that zebrafish even have MCH neurons, which are specialized cells that get activated right before the onset of the PMT wave and that are functionally identical to the MCH neurons that give rise to PGO waves in mammals.[32] In his recent book *The Hidden Spring: A Journey to the Source of Consciousness*, dream expert Mark Solms discusses the importance of PGO waves as drivers of human dream phenomenology. Leung and his collaborators are telling us that these waves do not make humans special since a version of them can be found in tiny fish that rarely reach four centimeters in length, fish separated from humans by more than 380 million years of evolutionary history.[33]

The takeaway here is that mammals and fish have remarkably similar sleep architectures despite having divergent evolutionary histories and dissimilar brain structures. Just as mammals have deep sleep and REM sleep, fish have SBS and PWS. The similarities between REM sleep in mammals and PWS in fish are astounding: in both cases, there is a distinctive neural signature, a unique brainwave that differentiates them from non-REM sleep and SBS, respectively; in both cases, this brainwave is set in motion by the activation of MCH neurons; and in both cases the brainwave starts in the pons and culminates in a whole brain state that registers "coherence indices" comparable to waking experience. These similarities support a phenomenological reading of PWS even in fish, especially given Leung's acknowledgement of the undeniably "wake-like character" of PWS itself.

Regrettably, Leung and his colleagues begin and close their article by claiming to be "agnostic" about the experiential correlates of the neural signatures they so meticulously describe. Apparently, they do not feel comfortable taking a stance on whether the phases of sleep they identified have a subjective component, whether they could be felt or experienced from the point of view of the fish. One cannot but wonder if they might even oppose the idea that zebrafish have their own subjective anchoring in the world, their own point of view.[34]

BEHAVIORAL EVIDENCE: THE PROMISE OF "ONEIRIC" ACTION

The French neuroscientist Michel Jouvet has convincingly argued that understanding dreams requires more than analyses of brain activity. It requires analyses of sleep-related behaviors. Behavioral evidence can elucidate whether an organism's sleep cycle is phasic or not, how an organism's body interacts with its brain during specific phases of sleep, and even how an organism might experience the dream state.

Video recordings of animal behaviors offer compelling evidence of dreaming. For example, there are endless online videos of sleeping animals behaving in ways that suggest dream experiences. YouTube is replete with videos of domestic companions acting out "waking-related displays" during sleep, including sleep-running, sleep-hunting, and sleep-mating. One of these videos, titled "Dog Dreaming," shows a dog sleeping quietly on their side. A few seconds later, two of their legs start twitching. After a few more seconds, the movement deepens as other limbs are slowly incorporated into a gradually rising behavior involving the entire body. Still horizontal and still asleep, the dog then breaks into a full-blown run as if chasing a target. The behavior is so well coordinated and executed that the dog eventually wakes up, stands up, and, in the confusion of it all, runs headfirst into the wall![35] The viewer sees a dog who is discombobulated as their immediate surroundings no longer match their expectations, which were undoubtedly set by the layout of the dreamscape.

Other videos show similar behaviors in cats, rats, rabbits, and octopuses.

Heidi's Dream, Revisited

The video of Heidi changing colors in her sleep that I discussed in the introduction may have captured the imagination of the general public when it first aired in 2019, but the truth is that not everyone was on board with Scheel's assertion that these changes probably reflected a dream experience.

Nicola Clayton and Alex Schnell, two animal intelligence experts from the University of Cambridge, were quoted in the *New York Times* arguing that the data does not support this conclusion. According to Clayton, we simply cannot know whether "the sequence of Heidi's color changes matches an experience she had while awake." To say that she is dreaming, as Scheel does, is "only conjecture." Her colleague, Schnell, concurs and reminds the reader that scientists have an obligation to opt for the simplest possible explanation of animal behavior—a methodological rule in animal research known as "Morgan's canon."[36] In this case, they argue, we do not need an explanation with cognitive or phenomenological components because a physiological one will do the trick just as well. Let's not say that Heidi was dreaming. Let's limit ourselves to what we know to be true, which is that the muscles controlling her color-changing organs were twitching.[37]

I agree with Clayton and Schnell that we need to be careful when interpreting the behaviors of animals, especially when dealing with one that is "the closest we may get to meeting an alien." But I disagree with their assumption that the simplest interpretation is always the most appropriate one. Caution, in and of itself, does not mandate that we shy away from explanations featuring cognitive, psychological, or phenomenological concepts. Yes, let's stick to the facts, but not without first asking what the relevant facts are that stand in need of explanation.

Recall that Heidi changed colors rather quickly—from a clean alabaster white to a yellow with patches of marigold orange to a deep purple bordering on midnight blue. Obviously, the muscles controlling her color-changing organs were contracting. Otherwise, there would have been no multicolored display. That is a fact. But there are other facts in play here as well. There is the unity of each display. Every display was sudden, global, and stable, and shared important similarities with a waking display. Then, there is the unity of the sequence of displays considered as a whole. This sequence mirrored the chain of behaviors that one would expect from an octopus hunting a crab in the middle of the day. These, too, are facts, and they, too, stand in need of an explanation. Why was each display so coherent? Why was the series so organized? As soon as we insist on the completeness of a purely physiological explanation, we bury these facts in the sand and give up on the possibility of a more cogent—though, yes, also more complex—explanation of the phenomena.[38]

The hypothesis that Heidi was dreaming of eating a crab may be conjecture, as Clayton says, but not in the everyday sense of a random guess that is no better or worse, epistemologically speaking, than a coin toss. It is conjecture in the sense of what philosophers of science call an "inference to the best explanation," which is a style of argument that first identifies the relevant facts and then selects the explanation that best accounts for all the facts combined. That this conjecture is fallible does not prove that it is not scientific. On the contrary, its fallibility is what makes it scientific in the first place.

Heidi's Cousin: The Cuttlefish

I admit that videos of sleeping animals are unlikely to settle theoretical disputes about animal mentation since they rarely control for confounding variables and are often open to competing interpretations. But when considered as part of a larger web of interlocking evidence, they lend support to Lindsay's nineteenth century claim that humans are not the only creatures on earth who dream. Even when considered in isolation, these videos matter since it would be hard to interpret the behaviors they capture as correlates of dreamless sleep. These behaviors are too well coordinated to be random, too coherent to lack a phenomenology. More than random motoric outputs, they appear to be intentional responses (or what the biologists Michael Chase and Francisco Morales call "integrated behaviors"[39]) on the part of the animal to situations that provoke them and relative to which they make good biological and psychological sense. We humans may never know what those situations are because we cannot step into the dream world of another animal. But that is beside the point. What matters is that such situations exist for the animal in question and that the behaviors they lead to—sleep-running, sleep-vocalizing, sleep-mating, sleep-chewing, and so on—are structured in such a way that it makes more sense for us to treat them as intentional responses to a concrete and meaningful situation than as brute reactions to external or internal stimuli.[40]

Still, video recordings can only take us so far. Luckily, they are not our only source of behavioral evidence of animal dreaming. Laboratory experiments have demonstrated that a broad range of species display the same "oneiric behaviors" that in humans we accept as trustworthy markers of dream experiences, usually during a phase of their sleep cycle that looks eerily like REM sleep, the phase of the human sleep cycle during which dreams are statistically most likely to occur.[41]

A good example of this research is a study on cuttlefish carried out by a group of scientists from the University of Pennsylvania in 2012. Like octopuses, cuttlefish are cephalopods with complex nervous systems and intricate chromatophoric systems of pigment-bearing cells that let them quickly camouflage themselves into the surrounding landscape. Led by Marcos Frank, an expert on sleep neuroscience, this research team was interested in whether these invertebrates sleep at all and whether their sleep cycle is uniform or phasic. To find answers, they introduced a group of cuttlefish into a "sleep chamber," which was an area in their tank that was optimized for resting, and recorded their behaviors over several days, tracking three variables along the way: whether the animals were active or inactive, whether their eyes were opened or closed, and whether their fins were moving or still.

From the collected data sets, they drew two main conclusions. The first was that cuttlefish undergo a "state of complete quiescence" that differs in observable ways from their waking state and is analogous to mammalian sleep. The second was that this state is not uniform but admits of two distinct phases: a phase of *absolute* quiescence that is devoid of all motoric output (the cuttlefish analogue of deep sleep) and a phase of *relative* quiescence that involves phasic motor activity including "twitching of the arms, eye movements, and nonrandom chromatophore activity" (the cuttlefish analogue of REM sleep).[42] During this second phase, "the eyes appeared to rapidly move beneath closed lids, chromatophore activity suddenly intensified, and the tips of the arms curled and twitched."[43] Given that the cuttlefish were asleep, Frank and his collaborators explain that these behaviors must have been "endogenous in origin, rather than exogenously driven by external stimuli."[44] That is to say, they must have come from the machinations of the cuttlefish's mind rather than the mandates of the external world.

Notably, the authors specify that the patterns of chromatophoric activity displayed by cuttlefish during the phase of relative quiescence "did not appear to be random firings of uncontrolled and uncoordinated neurons,"[45] a point that Louie and Wilson also make in their defense of rat dreams.[46] On the contrary, these patterns had a clear structure. During sleep, the cuttlefish exhibited the same "polarization of body patterning used in recognizing conspecifics."[47] In other words, they made the same split-body chromatophoric pattern during their version of REM sleep as when they ran into a familiar cuttlefish while awake.

I believe that these oneiric behaviors are too well integrated and mirror waking behaviors too closely for us to say they lack any subjective or phenomenological significance. Yet Frank and his colleagues are unwilling to acknowledge this significance, instead making a point of stating that their research does not support the hypothesis that cuttlefish dream. Their rationale, however, is different from that of Dave and Margoliash, who explicitly embrace an anti-phenomenological interpretation of sleep. For their part, Frank and his colleagues merely claim to be "agnostic" about the phenomenology of animal sleep. They argue that they cannot take a position on whether the chromatophoric activity displayed by the cuttlefish genuinely reflected "waking related displays" because they did not track all the relevant arousal variables in their study.[48] Without arousal data in hand, whether these animals experienced anything subjectively during their version of REM sleep is anybody's guess. We just do not know.

At first glance it may seem as if Frank and his colleagues are merely exercising due caution by not drawing unsupported conclusions, but their reluctance to speak about the phenomenology of cuttlefish sleep is suspect for two reasons. First, even if they did not track arousal variables, numerous other experts have. And their findings corroborate that animals experience sudden spikes in arterial pressure and heart rate during REM sleep. This is true of mammals such as humans, cats, dogs, and rats.[49] It is also true of fish. As fish fall asleep, their internal clock triggers a metabolic slowdown marked by diminishing cardiac and respiratory rhythms,[50] but there are also specific moments during their sleep cycle when the directionality of the metabolic process is reversed and they experience sudden but sustained spikes in cardiac and respiratory activity.[51] These spikes are likely physiological markers of a felt experience, and they may be good indicators of dreaming because they are typically coupled with non-physiological markers of dream sequences, as they occur "in close temporal association with hippocampal theta activity, PGO waves, and clusters of eye movements"[52]—all of which are customary gages of human dream phenomenology.

The second reason I am suspicious of this agnosticism about the dreams of cuttlefish is because Frank and his colleagues never considered the possibility that their findings may betoken parallels between sleep behaviors and waking displays *even in the absence of arousal data*.[53] Simply put, it would require something close to a statistical miracle for the chromatophoric system of a sleeping cuttlefish, which is composed of millions and millions of chromatophores, to regularly display controlled patterns of polarization by chance, especially when said patterns belong to the normal behavioral repertoire of the species and are associated with situations of clear evolutionary and social significance, such as the recognition of conspecifics.

If we look closely at their work, it quickly becomes apparent that Frank and his coauthors were aware of this problem but did not know how to handle it. This is why, immediately after claiming that it is "unlikely" that the sleep displays of cuttlefish mirrored their waking displays, they backpedal and concede that this possibility "cannot be entirely ruled out."[54] "It is therefore possible," they write, "that the nonrandom chromatophore activation observed during quiescence might be similar to patterns of activation that can occur during *vertebrate* REM sleep."[55] At one point, they even describe—as many other experts on animal sleep do—the sleep behaviors of the animals they study as "oneiric," which makes sense only if these behaviors are the result of a dream.

Chimpanzees "Talking" During Sleep

The last study I want to consider was conducted in 1995 by the primatologist Kimberly Mukobi, who developed an interest in the woefully understudied nighttime activities of captive chimpanzees.[56] Mukobi set out to investigate what Washoe, Moja, Tatu, Dar, and Loulis, five chimpanzees housed at the Chimpanzee and Human Communication Institute at Central Washington University, did at night once their human caretakers left and the lights went out.

After mounting five camcorders in the chimpanzees' enclosure and spending many nights observing the sleeping chimpanzees (and even more days watching and re-watching the more than 160 hours of footage she produced), Mukobi concluded that the drama of chimpanzee life does not cease when the sun goes down. It continues well into the wee hours of the morning. Chimpanzees continue their Machiavellian power struggles at night, probably because darkness offers the perfect cover for grooming powerful players, reinforcing existing friendships, and striking new allegiances while flying under the social radar. Even when they fall asleep, there is a clear social logic at work. Confirming Jane Goodall's observations at Gombe Stream Reserve during the 1960s, Mukobi found that chimpanzees are not indifferent to who they share their bed with at night. Most of them choose to sleep next to their best pals, as in the case of Dar and Loulis, who almost always slept "within an arm's reach of each other." More importantly for our purposes, Mukobi also documented several sleeping behaviors that suggest that our closest evolutionary cousins dream.

Several sleeping chimpanzees displayed "finger and hand twitching"[57] in the middle of the night, which Mukobi interpreted as evidence that they were "talking" in their dreams. Mukobi was able to go so quickly from observing the hand of a chimpanzee twitching to claiming they were talking in their dreams because, as part of their upbringing, all the chimpanzees in her study had been taught American Sign Language (ASL) to allow them to communicate with their human handlers and even with one another. Thus, the twitches Mukobi observed were signs—or, more specifically, *symbols*—made during sleep, the ASL-version of "talking" in one's sleep.[58]

Aside from the many partial ASL signs the chimpanzees made over the course of the experiment, Mukobi reports four instances of full-blown ASL signs, which clearly met the "PCM criteria" that primatologists use to identify, analyze, and interpret primate ASL communication. "PCM" stands for the *place* on the body where the sign is made (P), the *configuration* of the hands and fingers (C), and the *movement* of the hands and fingers during the act of signing (M). Here, it is worth quoting Mukobi at length:

> With regard to the four signs, Washoe made a sign that met the PCM criteria for COFFEE. The Place was on the inside (thumb side) of both hands, the Configuration was the right hand in the pincer position, and the left hand in a loose fist. The Movement consisted of the right pincer hand circling the left "C" hand, with both hands moving toward the ceiling at the same time. Loulis made hand movements on two separate occasions that met the PCM criteria for the sign for GOOD. On the first occasion, the Place was at his mouth, the Configuration was a relaxed "8" shaped left hand, and the Movement consisted of bringing the Configuration to the Place and tapping it twice. On the second occasion, the Place and Configuration were the same, but he used his right hand instead of his left. The movement consisted of the Configuration contacting the Place once instead of twice. Loulis also made a hand movement that met the PCM criteria for the sign for MORE. The Place was on

> the fingertips of the right hand, in front of his body. The Configuration was the left hand in a relaxed curve shape, and the Movement consisted of the left hand meeting the right hand fingertips, and contacting them several times in a pulse-like manner. The last observation was a gesture made by Dar. The Place of this gesture was in front of his body. The Configuration was the left hand in a loose "C" shape, and the Movement consisted of the Configuration moving from the body to the air space above his body (he was lying down), stopping briefly, and then falling down toward his body again. But because this did not meet the PCM criteria for any specific sign that Dar has acquired, it was not listed as a sign.[59]

For their behavior to count as a full-blown sign, the chimpanzees had to place their hands on the correct part of their body, make the correct finger configuration, and move their hands and fingers in accordance with convention, which is to say, in accordance with a recognized ASL rule. Washoe's sign for "coffee," to use my favorite example, involved putting both hands in front of the chest, making a different finger configuration with each hand, and coordinating the movement of both hands for a sustained period of time. The probability of this happening by pure chance is astronomically small.[60]

Now, the meaning of these behaviors is somewhat nebulous since Mukobi did not record the cerebral activity of the chimpanzees during the night and thus cannot say with

FIGURE 4. Captive chimpanzees "talk" in their sleep using ASL symbols. Here, Washoe makes the ASL sign for the word coffee, which consists of making a pincer with the right hand, making a C-shape with the left hand, and making the former circle the latter as both move away from the chest and toward the ceiling.

certainty that the behaviors occurred while the chimpanzees were in REM sleep. She writes, "these chimpanzees may have been talking, thinking, or even dreaming during their sleep."[61] But while she cannot guarantee that the chimpanzees were asleep, she is confident they were. "There were little clues such as deep breathing (sometimes snoring) and having their eyes closed." Besides, the overall context in which the signs took place was highly atypical:

When the chimps typically used signs (with the exception of Tatu, who would sign to herself while looking at magazines), they would gesture *toward* someone, usually their intended recipient. But in the nighttime cases, they were not directing their signs toward anyone. This, coupled with the fact that their eyes were closed and that they were breathing regularly, made me think they were sleeping.[62]

They were probably also dreaming, she says. Mukobi cites evidence that sleep talking is correlated with dreaming and that "sleep talking is not limited to individuals whose main form of communication is *spoken* language."[63] She explains:

> Signing during sleep has been reported in deaf humans (Raymond, 1990). In addition, Carskadon (1993) noted that finger movements may be another indication of talking and/or thinking during sleep in deaf people. The motor theory of thinking maintains that certain discrete muscular activities in the speech apparatus are closely associated with thinking. Carskadon used this theory to argue that discrete finger movements could be associated with thinking, as well. Max (1935) reported that sleeping deaf subjects showed much more finger electromyogram (EMG) activity than hearing people. He also found that the increased finger activity in deaf subjects was associated with reports of dreaming.[64]

If the sleep signs of human primates are indices of dreaming, why not those of nonhuman primates? Mukobi even reports watching Dar, one of the chimpanzees in her study, wake up kicking and pant hooting in the middle of the night. "One conclusion is that he heard a noise coming from outside of the building which woke him up and resulted in the display. However, no apparent strange noise could be heard on the video tape," she recounts. "Another conclusion could be that the display was an extension or 'acting out' of the end of a dream or nightmare."[65]

EVIDENCE FROM FUNCTIONAL NEUROANATOMY: LES CHATS DE JOUVET

A skeptic can always reason that even if other animals dream, their dreams may still differ from ours in philosophically important ways. Perhaps they lack the vividness, the high-resolution cinematographic quality, or the narrative structure of our dreams.[66] If so, one could grant that these animals experience a wide array of phenomenal states during sleep (such as seeing colors, smelling odors, or hearing sounds) while still refusing to call these experiences "dreams." For instance, if the nightly visions of animals turned out to be isolated phenomenal states that are not woven into a narrative sequence, one could argue that they are closer to the hypnagogic images we experience as we are falling asleep or to the hallucinations we experience when we succumb to heat exhaustion than to actual dreams.

Does such a qualitative difference exist between the dreams of humans and those of other animals? We obviously cannot ask animals about the contents of their dreams, so it is hard to say for sure whether their dreams are organized into a coherent narrative with a linear and causally linked progression of events and relationships. Yet research in functional neuroanatomy dating back to the 1960s suggests that this might be the case. More than cascades of unconnected phenomenal states, the dreams of animals seem to be action-packed sequences that fit into a clear narrative frame. As far as narrative structure goes, their dreams and ours may not be so different.

Earlier, I mentioned Michel Jouvet's claim that research into the neurophysiology of dreams needs to be reinforced by careful analyses of dream behavior. Jouvet is one of the most important dream researchers of the twentieth century and one of the few experts in the field who has no qualms about talking about the dreams of animals. He first became interested in dreaming in the 1950s, when it was already known that mammalian sleep is divided into periods of low and high cortical electroencephalogram (EEG) activity, with the latter correlating with observable rapid eye movements (REMs). However, the scientific consensus at the time was that REM sleep represented a form of "light sleep,"[67] during which nothing interesting goes on in the mind of the dreamer. As we fall asleep, our mind-bodies effectively shut down, meeting only the minimum biological demands needed to stay alive. While asleep, then, we are precariously balanced on the edge between the living and the dead. In this state, we do not have the luxury of performing such extravagant functions as cognition, intentionality, or conscious awareness. Each time we fall asleep, we effectively plunge into a mental abyss from which we somehow emerge upon awakening. This consensus prompted dream researchers in the 1950s to interpret the REMs that sleepers exhibit at regular intervals throughout the night as meaningless behavioral noise, random physical movements that sleep triggers for no apparent reason.

Jouvet rejected this view. He believed that REM sleep was not "light sleep" but "paradoxical sleep."[68] In a nutshell, the paradox was this: during REM sleep, the body is almost entirely passive (suggesting no underlying subjective experience), but the cortex is as active as during moments of waking perception (suggesting conscious awareness). This blend of low motor and high cortical activity needed a theory to explain it, and Jouvet claimed that we would only arrive at such as theory by taking seriously the possibility that REM sleep may be *dreamful* sleep, the phase of the sleep cycle during which, as Penelope laments in *The Odyssey*, we face "impossible, indiscernible" things.[69]

One of Jouvet's more thought-provoking claims was his assertion that the REMs we observe during paradoxical sleep are not "just an epiphenomenon reflecting anarchic variation of motor neurons," but "part of a structured, integrated motor behavior that is trapped somewhere in the nervous system."[70] During REM sleep, the sleeper mentally replays a unified behavioral program. This program is not expressed outwardly because sleep produces biochemical changes that induce a state of atonia in the sleeper, who subsequently loses voluntary motor control. These changes "lock" the behavioral program in the deepest recesses of the sleeper's mind. In most cases, REMs are the only components of this program that manage to escape these inhibitory processes and manifest externally. But REMs, Jouvet emphasizes, are only part of the program, just the tip of the iceberg.

In a series of experiments conducted in the 1950s and 1960s involving domestic cats, Jouvet set out to prove the existence of this hidden motor program by deactivating the neural mechanisms that inhibit its expression.[71] He believed that if he could only suppress the atonia-inducing mechanisms associated with REM sleep without compromising the integrity of REM sleep itself, he might liberate the trapped program, thus allowing sleepers to "act out" their dreams. To test this, he cut the dorsolateral part of the pontine reticular formation in a group of cats since research suggested that damage to this brain structure would suppress atonia but not REM sleep.

The results were astounding. When cats with pontine lesions entered REM sleep, they indeed "acted out" their dreams. They got up, meowed, walked around, groomed themselves, and explored their surroundings. They acted happy, angry, fearful, exploratory, and even sexually aroused.

Some of them stared intently into empty space as if stalking prey, ready to pounce, while others ran around their enclosures energetically fighting imaginary enemies like little furry Don Quixotes—all while fast asleep![72] These behaviors were so well executed, without the cats losing their balance or agility, that Jouvet said he could easily infer what each cat was dreaming about simply by comparing its behavior with a typical waking display. Is a cat gnashing his teeth while using "both front paws to try to capture some imaginary object"? He is probably dreaming of hunting. Is another taking quick and repeated paw strikes "into the void" with his "ears flattened backward and the mouth open ready to bite"? That one is probably dreaming of having a fight.[73]

Jouvet underscores that, when considered in relation to the cats' objective surroundings, these behaviors served no obvious purpose since, in the laboratory, there was no real prey to catch, no real enemy to vanquish. These behaviors appear functional and purposeful only when considered in relation to the cats' *dream* worlds and to the intense feline drama that must have unfolded in them. For this reason, these behaviors point to a subjectively constituted reality that existed solely in the cats' minds—a reality that is clearly organized as a narrative. As I read it, Jouvet's research demonstrates that animals undergo complex lived experiences during sleep that are more than haphazard aggregations of raw sensations. They are unified perceptual realities—"visual scenarios," as the anthropologist Derek

FIGURE 5. A cat in Michel Jouvet's laboratory fights an imaginary enemy after the neurons responsible for muscle atonia, which are located in the pons, have been surgically removed. Sadly, the cat shows all the behavioral signs of distress and anxiety, such as opening the mouth to bite, turning the ears back and flattening them, and striking the air with his front paws.

Brereton says[74]—in which events follow in sequence, supported by a narrative arc.

THE EVIDENCE IN PERSPECTIVE

Presently, it is hard to say exactly how widespread dream experiences are in the animal kingdom because we know a lot about the sleep patterns of some animals but next to nothing about others. Also, what counts as evidence of dreaming remains an open question, especially in the case of species other than *Homo sapiens.*

In this chapter we have looked at empirical evidence rooted in electrophysiological, behavioral, and neuroanatomical research, but even this evidence poses conceptual challenges. Electrophysiological data is often open to competing interpretations. Drawing conclusions about subjective experience based on it can be tricky business, especially when the same subjective experience can be achieved by vastly different brain states. Humans, for instance, can experience dream imagery while falling asleep, during REM and non-REM sleep, and even during the waking state, although brain activity is not identical across these states.[75] Similarly, behavioral data can be inconclusive. On the one hand, some sleep-related behaviors are not reliable indices of dream experiences, and assuming otherwise can lead to false positives. On the other hand, we are in the habit of expecting the oneiric behaviors of other animals to mirror ours, when we should expect them only to make sense relative to the animals themselves (such as rapid whisker movement in cats). It is likely that this expectation has already led to a barrage of false negatives.[76] Finally, while neuroanatomical data has played a central role in debates about cognition, we are not sure how much weight neuroanatomical similarities should be given, especially when organisms can perform similar cognitive functions despite having dissimilar neural organizations.

These challenges are real, but not insuperable. In isolation, electrophysiological, behavioral, and neuroanatomical findings may not be entirely compelling, but together they form a robust evidentiary network that supports the following conclusions about animal dreams:

1. Many animals experience a state of sleep analogous to what Michel Jouvet calls "paradoxical sleep," which is the stage of sleep during which humans typically dream.
2. During this stage of sleep, animals mentally replay waking displays.
3. Replay often unfolds on the same timescale as the waking behavior it recapitulates.
4. This replay has an arousal profile (changes in heartbeat, respiration, blood pressure, etc.).
5. Replay tends to be coupled with oneiric behaviors that require the coordination of the animal's entire motoric system (such as sleep-running) or parts of it (such as REMs).
6. Not all the behaviors associated with a dream are expressed under normal sleep conditions, but they can be liberated by deactivating the brain processes that inhibit their expression.
7. Oneiric behaviors are best interpreted as responses to lived situations that are biologically and psychologically meaningful for the animals involved, rather than as mechanical reactions to physical stimuli.

An Impossible Line in the Sand: Starting from Mammals

None of this settles the issue of which animals dream, an exceedingly difficult question to answer. As things currently stand, the case for mammals is the strongest. In an article published in 2020 in *The Journal of Comparative Neurology,* Paul Manger and Jerome Siegel argue that this case is so strong that the question we ought to be asking nowadays is whether there are any mammals who do *not* dream (a question that they answer affirmatively). But, as they are quick to point out, the answer depends on whether we take a "hard" or a "soft" stance about the relationship between dreaming and REM sleep. The hard stance holds that dreaming occurs only during REM sleep, while the soft one maintains that it can occur during REM or non-REM sleep. If we embrace the hard stance, some mammals will not make the cut (monotremes, cetaceans, and pinnipeds) and some will be borderline cases (African elephants, Arabian oryxes, rock hyraxes, and manatees).[77] If we embrace the soft stance, the only mammals excluded may be cetaceans, the only biological order whose sleep phenotype seems logically incompatible with dreaming of any kind. "Cetaceans," they say, "appear to be the least likely of all the mammalian species to experience any form of mental representation during sleep that could be readily defined as dreaming."[78]

I will not try to persuade anyone to take either of these stances because regardless of which we adopt, we cannot avoid an affirmative answer to Manger and Siegel's titular question

("Do All Mammals Dream?"). With a few possible exceptions, all mammals dream. These exceptions surely cannot be overlooked, but they seem insignificant in the grand scheme of things, especially when we keep in mind that an article in the *Journal of Mammalogy* recently estimated that there are more than six thousand mammalian species in existence.[79]

And mammals are but one branch of the animal kingdom. If we look at the electrophysiological data, it is likely that birds and fish also dream. Behavioral data shows that animals of all classes exhibit oneiric behaviors. This includes: mammals, such as mice, rats, rabbits, dogs, cats, chimpanzees,[80] opossums,[81] platypuses,[82] echidnas,[83] squirrel monkeys,[84] and beluga whales;[85] birds, such as zebra finches,[86] ostriches,[87] penguins,[88] owls,[89] pigeons,[90] vultures,[91] and chickens;[92] reptiles, such as Australian dragons,[93] chameleons,[94] iguanas,[95] and lizards[96] (though the jury is still out on crocodiles[97] and turtles[98]); and cephalopods, such as cuttlefish[99] and octopuses.[100] The discovery of oneiric behaviors in this last group is particularly startling, for it implies that dreaming may have evolved independently in at least two phyla (*Chordata* and *Mollusca*). If true, this would have colossal implications for contemporary dream research. It would refute Michel Jouvet's hypothesis that paradoxical sleep is limited to homeotherms (birds and mammals),[101] as well as the biologist Ida Karmanova's more radical but less popular hypothesis that it exists only in homeotherms and poikilotherms (fish, amphibians, and reptiles).[102] If cephalopods can dream, then endogenous dream sequences must be much more widespread throughout the animal kingdom than previously imagined, extending across seemingly impassable evolutionary distances.[103]

Even so, the unavoidable fact is that we eventually hit a limit. We may not hesitate to say that mammals and birds dream, as do octopuses. But as we move across the endless branches of the tree of life, the dream hypothesis slowly comes undone at the seams. Can chimpanzees dream? Yes. What about octopuses? Yes. Can we extend it to fish? Probably. What about ants, bees, and sea sponges? Suddenly, continuity gives way to discontinuity as we sense that we have crossed a line along the way, even if we cannot say exactly when or where. The physician Andrew Freiberg put his finger on this problem when he wrote, "Dreams may be an essential function of sleep in humans and other primates or even all mammals, but extending that function to earthworms and daylilies is difficult to imagine."[104]

I do not pretend to solve this problem in this book. How could I, when the main lesson from biology in the one hundred and fifty years since the publication of Darwin's *On the Origin of Species* is that in nature there are no perfect lines? But let's not lose our sense of perspective. Even if we do not have the comfort of a sharp line distinguishing the animals who dream from those who do not, we find ourselves on very different terrain from where we started: miles from the view that only mammals dream and lightyears from the view that *Homo sapiens* are the only dreaming animals on earth. A choice looms for us: either we hew to an anthropocentric theory of dreaming and ignore the scientific findings presented in this chapter, or we follow the life sciences into the uncanny world of animal dreams. We may not love the options on the table, but the middle ground between them is quickly disappearing.

Minding the Gap

We must not confuse the claim that we can make educated judgments about whether animals dream with the assertion that we have unlimited access to the contents of their dreams. As my analysis of animal emotion in chapter 2 will show, sometimes we have partial access to this content, but this access is always limited and never without problems.

As a rule of thumb, analyses of animal dreams must be guided by two overarching principles. One is the principle of interspecies differences, which encourages us to approach the issue on a species-by-species basis, paying close attention to the different sleep cycles, perceptual systems, cognitive abilities, and evolutionary histories of each species.[105] The other is the principle of intraspecies variation, which recognizes vast differences in the sensory, physical, and cognitive capacities of different members of the same species.[106] These principles drive home the point that even if we have some access to the dream worlds of other animals, we must acknowledge the limits of this access and respect the differences that distinguish one species from another and one organism from the next. Each dream world is *theriomorphic* (from the Greek *therio,* meaning "beast" or "animal," and *morph,* meaning "form" or "shape"). It takes the "form" of the specific animal whose world it is.

This, I admit, leaves us in a discomfiting epistemic position—familiar yet alien, proximate yet remote. But we need to learn to sit with this discomfort, because it is in this liminal space that new possibilities open between us and other animals, including the possibility of discovering something meaningful about those other spirits who roam the world on their own but by our side. Indeed, I would even say that more than learn to sit with this discomfort, we must learn to cherish it. After all, we will never fully capture animals in our conceptual, linguistic, and hermeneutic nets. The best we can do is strive to understand what connects us while respecting the many things that separate us.

Dreaming illustrates this polarity well. Other than anthropocentric conceit, we have no reason to expect other animals to dream in the exact same way we do. In the waking state, the worlds other animals inhabit are already radically distinct from ours, since they depend on those animals' sensory modalities, motoric possibilities, and ecological affordances.[107] Can we expect their dream worlds to be any less divergent, any less alien, any less nonhuman? We know, for instance, that humans rarely report smells in their dreams. But given the centrality of smell in the world of dogs, dog dreams may be more olfactory than visual. Likewise, zebra finch dreams may be sonorous experiences without visual or olfactory content. These differences would not disqualify these experiences as dreams; if anything, they would qualify them as "olfactory" and "musical" dreams, respectively. The philosopher Ludwig Wittgenstein famously said, "If a lion could talk, we could not understand him."[108] And if a lion could dream—and I suspect they can—we would not be much better off. We might know that he dreams, but not what his dream ultimately means *to him.*[109]

RECLAIMING OLD WISDOM

I conclude with two brief observations. The first is that the above considerations invite us to rediscover the wisdom of nineteenth century naturalists who looked for signs of complex mental states in animals and openly "accredited" them, to use George Romanes's term, with the capacity to dream.[110] Instead of dismissing their theories as the phantasmagoric illusions of an uncritical mind, we should ask ourselves whether these theories might become more rather than less appealing with the passage of time. In the history of science there are endless examples of new discoveries extending unexpected lines of credit to theories previously thought obsolete, giving them a new lease on life. As the French philosopher Gaston Bachelard once said, "With any significant change to the layout of contemporary knowledge, old names suddenly become important again."[111] As we embark on this voyage of rediscovery, we must simply ensure that we rekindle old ideas with the light of present-day knowledge, interests, and concerns.

The second is that our analysis of animal sleep research underscores the point that science is never free of philosophy. Scientific research is always haunted by questions about the meaning of data that will never be answered with more data. As such, we must calibrate our best science with our best philosophy. In this case, this means enriching what we know scientifically about animal sleep and animal cognition with what we know philosophically about the nature of dreaming and consciousness.

To this task we now turn.

CHAPTER 2

Animal Dreams and Consciousness

> Animals dream. Am I altogether in error in thinking that the philosophical and historical implications of this platitude are momentous, and that they have received remarkably little attention?
>
> —GEORGE STEINER[1]

A PHILOSOPHICAL MONSTER

Although there is an abundance of arguments in favor of animal consciousness in the neuroscientific, psychological, and philosophical literatures, they all have one thing in common: they pay no attention to what the minds of animals do during sleep.[2] All of them focus on what animals do when they are awake, alert, and actively engaged with their surroundings. Can they feel pain and pleasure? Can they understand the intentions of others? Can they experience emotions like joy, empathy, or grief? Can they solve puzzles, understand inferences, or grasp abstract concepts? This focus is understandable, since eliciting, controlling, and interpreting the waking behaviors of animals is much easier than studying whatever they do while asleep. But, in limiting ourselves to waking behaviors, is it possible that we are missing an entire dimension of animal experience that might enrich our understanding of animal minds? Might we be inadvertently turning our backs on a vector of animal consciousness whose philosophical implications are, as Steiner says, "momentous"? I believe we are.

In this chapter, I develop a case for animal consciousness that centers on the dreams of animals. My thesis is that it is impossible for an organism to dream and to lack consciousness. Since we have it on good authority that many animals dream, it follows that those animals must be conscious agents with their own perspective on the world—even if that perspective, like Heidi's or like Wittgenstein's lion's, is arrantly unlike ours. The very idea of a nonconscious dreamer amounts to what Wittgenstein himself would call a "philosophical monster," a notion so absurd that no respectable philosophical theory could embrace it.

I present this case in two steps. To start, I demonstrate that dreaming is a sufficient but not necessary condition for consciousness, understood very broadly as the property of being aware (as opposed to being nonconscious).[3] Once this is established, however, I leave this broad understanding of consciousness behind and follow contemporary scientists and philosophers in thinking of consciousness as a complex phenomenon that comes in different *types* rather than a monolithic property that is an all-or-nothing affair. Thus, in the second stage of the argument, instead of arguing that animals possess a singular thing called "consciousness," I consider the specific types of conscious awareness that we can meaningfully attribute to them. To this end, I present a novel model of consciousness—called the SAM model—that segregates consciousness into three types: *subjective*, *affective*, and *metacognitive*. These terms will be explained below but, briefly, they refer to whether animals experience themselves as the center of their own phenomenal world, whether they experience affects, feelings, and emotions, and whether they are able to monitor their own mental states. Here, I connect this model of consciousness to animal dreaming by demonstrating that the dreams of animals are *always* evidence of subjective consciousness, *often* evidence of affective consciousness, and *occasionally* maybe even evidence of metacognitive consciousness.

DREAMING AS A SUFFICIENT CONDITION FOR CONSCIOUSNESS

The idea that dreaming entails consciousness is not new. In the 1980s the psychologist David Foulkes held that we do not achieve consciousness because we dream; rather, "we dream *because* we have achieved consciousness."[4] Throughout the 1990s, several influential philosophers and neuroscientists echoed this view. John Searle defined dreams as "a form of consciousness, though in many respects they are quite unlike normal waking states,"[5] while Paul Churchland, the foremost defender of eliminative materialism, claimed that "the sort of consciousness one has during dreams is decidedly nonstandard, but it does appear to constitute another instance of the same phenomenon."[6] Like Foulkes, Searle and Churchland see dreaming, by definition, as a mode of conscious awareness.

The philosopher Evan Thompson develops this position in his book *Waking, Dreaming, Being: Self and Consciousness in Neuroscience, Meditation, and Philosophy*. According to him, Western philosophers have historically treated consciousness as a binary property that is either wholly present or wholly absent in an organism at any moment, like a light that turns itself on and off. On this classical conception, we are conscious only when we are awake, alert, and in full possession of our mental faculties. At all other times, including when we are dreaming, we are unconscious or nonconscious.[7]

Drawing upon ancient Indian yogic traditions, Thompson encourages us to abandon this binary conception on the grounds that consciousness is a multimodal phenomenon that takes different forms at different times. More than a light that switches between two states, consciousness is a switchboard whose configuration at any moment in time depends on a number of biological, physiological, psychological, and even social variables. Based on his reading of the "Great Forest Teaching" of the *Upanishads*, a group of texts written around the seventh century BCE that express the core ideas of Hinduism, Thompson identifies four modes of consciousness, which he describes as waking consciousness,

TABLE 1. *Evan Thompson's Theory of Consciousness*

Waking Consciousness	Dreaming Consciousness	Dreamless Sleep	Pure Awareness
The state of being awake and attentive to one's surroundings. It implies the ability to focus on specific aspects of one's phenomenal field.	The state of being phenomenally conscious during sleep, especially REM sleep. During dreams, one is attentive to events in the dream world.	The state of being asleep but without dreaming, usually during non-REM sleep. For Buddhists, this state remains conscious, but in a minimal way.	A controversial state associated with heightened insight attained during meditative practice and at the moment of death.

dreaming consciousness, dreamless sleep, and pure awareness (see Table 1).

While these modalities differ from one another phenomenologically, they share what Indian and later Buddhist texts identify as the hallmark of all conscious awareness, which is *luminosity*. They all "illuminate" (or, as Western phenomenologists might say, "disclose") phenomena for an observer. They make the world "show up" for a subject:

> "Luminous" means having the power to reveal, like a light. Without the sun, our world would be veiled in darkness, but without consciousness, nothing could appear. Consciousness is fundamentally that which reveals or makes manifest because it is the crucial precondition for appearance. Nothing, strictly speaking, *appears* unless

> it appears *to* some consciousness. Without consciousness, the world can't appear to perception, the past can't appear to memory, and the future can't appear to hope or anticipation.[8]

Consciousness is "that which makes something manifest and apprehends it in some way," that which lights up a field of perception that a subject immediately grasps and experiences as *theirs*.[9] Thompson claims that since these modes share this luminous quality equally, they are all necessarily conscious even if our experience of some of them is, as Churchland would say, "decidedly nonstandard."

Two of the world's leading experts in the philosophy of dreaming, Jennifer Windt and Thomas Metzinger, reach the same conclusion by a different route. Instead of turning to theories of consciousness rooted in ancient yogic traditions, they look at the mode of conscious experience that Western philosophers have always privileged (namely, waking experience) and wonder whether the conditions that make it "conscious" might also be met in other modes, especially dreaming. Waking experience, they say, is conscious because it satisfies three formal constraints:

1. Presentationality, which entails "the presence of a world" that represents a here and now. This world makes itself manifest to a subject whose consciousness is always consciousness *of* some aspect of this world.

2. Globality, which entails "the activation of a global model of reality" that engulfs the subject in their totality. There can be no external perspective, no God's eye view, from which the subject could, even in theory, grasp this model as one object among many since this model envelops the subject and establishes the parameters of their reality.
3. Transparency, which presupposes that the subject experiences their global model of reality *as reality itself* and not as a model of it. The model's status as model must be inaccessible to the experiential subject, for whom the model must remain "transparent."[10]

To make these constraints more concrete, here is a simple example. When I run down the street to catch the bus, I am "conscious" insofar as I am immersed in an all-encompassing world that is present-to-me, a world that I cannot grasp from an external point of view. As far as I am concerned, this world is the real here and now. It is not a simulation. It is not a figment of my imagination. When I run to catch the bus, I am "conscious" because I am presented with a global and transparent reality. But if the reason we describe this experience as "conscious" is because it meets these criteria, there is no justification for describing dream states any differently. When I *dream* of running to catch the bus, I am also immersed in a here and now; I also experience the here and now of my dream as a totality that envelops me rather than as a spectacle that I gaze at from a distance; and I also experience this totality as real rather than ersatz. When I dream, therefore, I must be conscious, even if I am not necessarily conscious of being in a dream.[11] In my dreams, I am not conscious *of* dreaming; I am conscious *as* dreaming:

> From a purely phenomenological perspective, dreams are simply *the presence of a world.* On the level of subjective experience, the dream world is experienced as representing the here and the now. And even though it is a model constructed by the dreaming brain, it is not recognized as model but is experienced as reality itself. Put in philosophical terms, one can say that the reality model created by the dreaming brain is philosophically transparent; the fact that it is a model is invisible to the experiential subject.[12]

Windt and Metzinger then add: "one can say that dreams are conscious experiences because they satisfy [these] constraints."[13] This reiterates the point Foulkes made in the 1980s that dreaming logically entails consciousness because when we dream, we stand before a phenomenal field that presents us with an entire world. Adjusting Descartes's famous seventeenth century refrain, we might say: "I dream, therefore I am."

THE SAM MODEL OF CONSCIOUSNESS

Consciousness is notoriously hard to define. Experts have not reached a consensus about what it means, what its antonym is, how it comes into being, or even how we can tell whether someone has it at any point in time. Irked by this difficulty, the psychologist George Miller famously called for a moratorium on the use of this term in scientific writing in the early 1960s as he feared that a term with such loose contours would only muddy the otherwise limpid waters of psychological research:

> Consciousness is a word worn smooth by a million tongues. Depending upon the figure of speech chosen it is a state of being, a substance, a process, a place, an epiphenomenon, an emergent aspect of matter, or the only true reality. Maybe we should ban the word for a decade or two until we can develop more precise terms.[14]

Miller's injunction was well received by his contemporaries, and a *de facto* moratorium was put in place. For decades, endless books were published about the human psyche without any reference whatsoever to "consciousness" or "conscious experience." In *Le Code de la Conscience*, the French neuroscientist Stanislaus Dehaene, known for his version of the Global Workspace Theory of consciousness, recounts how anything resembling consciousness-talk was given a wide berth in scientific articles, journals, and conferences well into the 1980s. Young researchers who entered the field during this period were encouraged to concoct experiments to test how, when, and why people become conscious of different stimuli and to publish their findings in the most prestigious journals without ever mentioning the supposed "consciousness" of their subjects. It was thought, says Dehaene, that the concept "didn't bring anything fundamental to scientific psychology."[15] It was considered superfluous.

Today, consciousness is no longer considered off-limits for scientists, but the problem of how it should be defined lingers. Over the past couple of decades, a loose consensus has formed in philosophy of mind, cognitive psychology, and neuroscience that the best way to tackle this problem is by dividing consciousness into smaller categories and seeking to understand how they operate and interact with one another. Nowadays, it is quite common for experts interested in the topic to begin from the assumption that consciousness comes in various forms, each even admitting of varying degrees.

One of the benefits of this approach is that it makes it easier to grapple with the malleability, versatility, and multidimensionality of conscious experience without having to settle upon a grandiose definition of consciousness ahead of time. One of its drawbacks is that it has led to a proliferation of taxonomic schemes, each of which tries to carve at the joints of consciousness in its own way.[16] For the purposes of this book, and at the risk of adding yet another brushstroke to an already crowded canvas, I propose that we divide consciousness into three subspecies of awareness: subjective, affective, and metacognitive. Subjective consciousness refers to being the center of one's phenomenal reality; affective consciousness refers to emotions, feelings, and affects; and metacognitive consciousness refers to forms of cognitive processing that involve some kind of reflexivity. For short, I call this the SAM model of consciousness—with "S" for subjective, "A" for affective, and "M" for metacognitive (see figure 6).

I have settled on this triad because these categories appear with remarkable frequency, under one guise or another, in contemporary research about human dreams, and I want to see how far they extend across species lines. My goal in presenting this model is not to offer a complete picture of conscious experience across the entire animal kingdom since

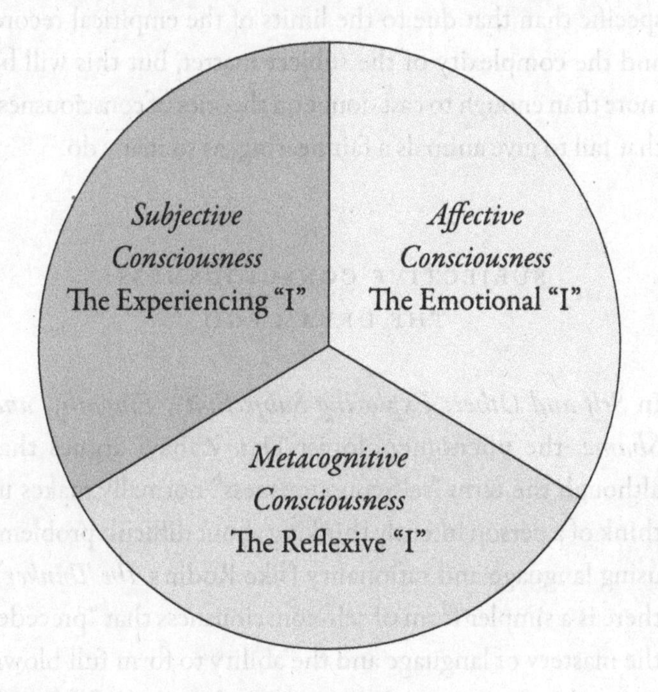

FIGURE 6. The SAM model of consciousness.

I do not believe that is possible. My goal is also not to imply that subjectivity, affect, and metacognition exist in isolation of one another since I believe they intertwine in highly complex ways. My goal is less ambitious: to show that we can thicken our understanding of animal consciousness by zeroing in on these three categories and clarifying their relationship to dreaming.

In a nutshell, I submit that *all* animals who dream are subjectively consciousness, that *many* are also affectively consciousness, and that a *handful* may even have metacognition on top of that.[17] As mentioned before, it is hard to get more specific than that due to the limits of the empirical record and the complexity of the subject matter, but this will be more than enough to cast doubt on theories of consciousness that fail to give animals a fair hearing, as so many do.

SUBJECTIVE CONSCIOUSNESS: THE DREAM EGO

In *Self and Other: Exploring Subjectivity, Empathy, and Shame,* the phenomenologist Dan Zahavi argues that although the term "self-consciousness" normally makes us think of a person intently thinking about difficult problems using language and rationality (like Rodin's *The Thinker*), there is a simpler form of self-consciousness that "precedes the mastery of language and the ability to form full-blown rational judgments and propositional attitudes."[18] Building on the writings of French and German philosophers such as Edmund Husserl, Jean-Paul Sartre, and Maurice Merleau-Ponty, Zahavi describes this primary form of consciousness as "the ongoing first-personal manifestation of one's own experiential life."[19] Well before we develop the cognitive functions that underpin rational introspection, we develop a basic form of self-awareness that is built into the frame of our lived experience, that *is* the frame of our lived experience.

I interpret this primal form of experience, which I call subjective consciousness, as entailing two things:

1. a feeling of subjective presence, which is the feeling that one exists at the epicenter of one's world and subsists in this position for an extended period of time; and
2. a sense of bodily self-awareness, which is the feeling of having a tacit understanding of one's own body. This, in the words of the bioethicist David DeGrazia, is "awareness of one's own body as importantly different from the rest of the environment."[20]

Animals who have (1) and (2) are subjectively conscious and have at least a minimal sense of self.[21] In this section, I want to connect this account of subjective consciousness to dreaming by showing that in the dream world we always have a feeling of subjective presence and a sense of bodily awareness.

Subjective Presence: The Manifestation of the Dream World

In "The Immersive Spatiotemporal Hallucination Model of Dreaming," Jennifer Windt argues that all dreams share a "phenomenological core," which is the dream ego's experience of being immersed in a spatiotemporally organized reality.[22] According to her, no matter how unstable, illogical, or weird it may be, every dream is organized around a dream ego who is "there" in the dream world and with whom the dreamer ultimately identifies. It is this ego who experiences the dream as a lived reality, who suffers the highs and the lows of the dream, and who has panoptical vision in the dream itself.[23] It is this ego, too, who subsists from the beginning to the end of the dream:

> The crucial factor that distinguishes dreaming from non-dreaming sleep experiences is precisely *the sense of spatial and temporal presence in the dream*. In a very basic sense, there is a hallucinatory scene that is organized around an internal, spatiotemporal first-person perspective as well as a sense of spatiotemporal self-location, i.e., the sense of occupying a space (even a point will be extended in a minimal sense), plus an experienced "now" and the experience of duration.[24]

For Windt, the key term is "presence." Without this subjective sense of presence, there would be no dream because dreams are inherently egocentric experiences that revolve around a subjective axis. This axis is what makes us feel that we are "there" when we dream and that the events happening in the dream are happening *to us.* Windt concludes that any creature who dreams is necessarily endowed with what she calls "minimal phenomenal selfhood," a basic form of subjectivity that hinges on nothing more than occupying a location in a spatiotemporal field and subsisting in it for an extended period of time.[25] Only beings endowed with this minimal phenomenal selfhood can, in principle, dream.

Writing from a slightly different angle, Evan Thompson elucidates the subjective tethering of dreams by appropriating the neurologist James Austin's concept of the "I-Me-Mine" structure of waking experience. He explains:

> Ordinary waking consciousness is conditioned by what Austin, in *Zen and the Brain,* calls the "I-Me-Mine." "I" is the self as thinker, feeler, and actor; "Me" is the self as affected and acted upon; "Mine" is the self as possessor, the appropriator of thoughts, feelings, body characteristics, personality traits, and material possessions. This tightly knit and mutually reinforcing triad constitutes our sense of the self as ego, our deep-seated impression of being a distinct and bounded self standing over and against the world.[26]

While the "I-Me-Mine" is not a universal norm of human life (given that it vanishes in dreamless sleep and certain comatose states), it is a universal norm of waking experience that is systematically recreated in dreams. "Dreams typically recreate this structure, for often one has a dream body in which one participates in the dream world. Even when one experiences oneself as an observational point of view, one still experiences oneself as a subject situated in relation to the dreamscape."[27] "When we dream," Thompson goes on to say, "we experience being in a dream; more precisely, we experience being in the dream world. The experience of being a self in the world, which marks the waking state [. . .], reappears in dreams."[28]

Bodily Awareness: A Theory of Embodied Dreaming

Windt's and Thompson's analyses underscore the feeling of subjective presence that accompanies all dream experiences. But subjective consciousness, as I have defined it, requires a sense of *bodily* self-awareness as well. Given that our physical bodies are generally immobilized during sleep (as we saw in chapter 1), is there any sense in which we are "embodied" in dreams? According to some prominent dream scientists, the answer is yes because the dream ego, that subjective point around which all dream sequences pivot, is always already "phenomenally embodied."[29] At each moment of a dream, our dream ego occupies a specific spatiotemporal location in the dreamscape and has a dream body with a specific size and shape.

The French existentialist Jean-Paul Sartre understood this already in the 1940s. In *The Imaginary: A Phenomenological Psychology of the Imagination*, he explains that "the person of the dream is always somewhere [. . .], and this 'somewhere' is itself situated in relation to a whole world that is not seen but is all around."[30] The phrase "all around" is crucial since, for Sartre, the dream world is the circumference of our dream experience, and our dream body is its center. Sartre talks about the dreamer as having a "body schema" that is different from, but not necessarily inferior to, the body schema of the waking ego. This body schema determines what the dream ego can do and enables it to differentiate on a pre-conceptual level between self and other, between "I" and "not-I."[31]

Sixty years later, cognitive neuroscientist and eminent dream theorist Antti Revonsuo would make the same argument. According to him, every dream ego has a unique physical location and a determinate body image, and the two are intricately interconnected. The dream ego possesses a body image *because* it has "a bodily existence and location in the dream world." In other words, to experience oneself as having a bodily existence and occupying a specific location in the world is enough to produce a body image. "In this respect, the dream self is not all that different from the waking self," he says.[32] This idea appears once more in the writings of the dream expert Derek Brereton, who judges that the "self-as-body-image is phenomenologically intrinsic to dreaming." There is no dream without a dream ego, no dream ego without a dream body, and no dream body without a dream body-image. Brereton

FIGURE 7. The philosopher Jean-Paul Sartre and the neuroscientist Antti Revonsuo believe that all dreams, even those that seem utterly incoherent, abide by a clear subjective logic since they all involve an embodied dream ego at the center of the dream world.

is thus led to describe this body-image as "the primal gestalt" of our dreams.[33]

The embodiment of the dream ego implies that the dreamscape is more than a homogenous sensorium in which an abstract and disembodied ego dwells; rather, it is a field of subjective paths and openings that disclose a world for a subject.

These paths and openings inundate our perceptual field with lines of tension that incline us to pursue some projects rather than others, to actualize some possibilities rather than others. This is the reason Sartre makes the same claim about the dream world that experts on embodied cognition such as Maurice Merleau-Ponty, Francisco Varela, Eleanor Rosch, Alva Noë, and Dan Zahavi make about the waking world—namely, that it is not a boundless, three-dimensional Newtonian expanse that is indifferent to our bodies, interests, or goals. It is a "hodological space"[34] (from the Greek *hodos,* meaning "path") that is coextensive with the range of our action potentials.

These potentials do not exist independently of our body—or, rather, of our dream body. They are functions of its structure, position, and orientation and have meaning only in relation to it. In its dynamic interaction with its environment (that is, the dreamscape), this body determines what counts as a possibility or as a limit, as an aperture or as a barrier for us. The sum of these possibilities and limits gives the dream world its parameters, transforming our dreams into subjective realities that we live through with the full force of our being more than endogenous moving pictures that we regard with the passive disinterest of a wearied spectator.

The Subjective Scaffolding of Dreams

Dreams, then, are subjectively structured. They require the presence of an ego endowed with a feeling of subjective presence and a sense of bodily self-awareness. Even when experts quarrel about the extent to which we exert bodily agency in our dreams or the degree to which we identify with our dream ego, it is understood that a dream purged of all subjective structuring would be no dream all. Even seemingly egoless dreams, such as the dreams of young children or the dreamlike images we experience while we are falling asleep, turn out to have a subjective anchor without which they would not be recognizable as dreams.[35] Subjectivity is dreaming's primary condition of possibility. This, Sartre says, is why we cannot witness the death of our dream ego in our dreams: the death of this ego would bring about the instant and irreversible dissolution of the dream world itself. Death, the philosopher Martin Heidegger famously said, always eludes us—to which Sartre would add, even in our dreams.

This ontological link between dreaming and subjectivity means that there can be no dream without an ego. It is hard to even imagine what such a dream would look or feel like. In an egoless dream, who would be dreaming? From what perspective would the dream be experienced? Who could claim it as theirs? An egoless dream is phenomenologically impossible because where there is a dream, there must also be an ego that actualizes, sustains, and experiences it, an ego that is the ultimate base of its existence. An egoless dream is as inconceivable as a shapeless statue or an invisible painting, as paradoxical as a sun that does not shine or a river that does not flow.

Importantly for us, this ego need not be human. Any creature who dreams is necessarily subjectively conscious and therefore has a feeling of subjective presence as well as a sense of bodily self-awareness. Creatures who dream must be subjects for whom a world of experience "lights up" and not, as Peter Godfrey-Smith puts it, "biochemical machines for which all is dark inside."[36]

AFFECTIVE CONSCIOUSNESS: THE EMOTIONAL HORIZON OF THE DREAM WORLD

The Origins of Modern Dream Science: Rebelling against Freud

During the golden age of dream research that lasted from the 1950s to the 1970s, scientists believed that dreams were created by random activity in the pons, a part of the brainstem that connects the medulla to the midbrain. At the heart of the "pons activation hypothesis" that dominated this period was the idea that dreams are nothing more than physiological white noise produced by the random firing of neurons in the pons.[37] In this regard, dreams were thought to be similar to, say, the sound our blood makes as it races through the circulatory system. In both cases, we may identify the organic causes of the phenomenon, but the moment we try to decipher its "meaning" we are led on a wild-goose chase. Like the sound of blood, our dreams are fundamentally meaningless and therefore uninterpretable.

Interestingly, this hypothesis became fashionable in the 1950s less due to its empirical credentials and more because it gave dream scientists ammunition against Sigmund Freud's psychoanalytic theory of dreams, which had enjoyed half a century of success in Europe and North America, but which new generations of scientists began rejecting as pseudoscientific mumbo-jumbo. At the dawn of the twentieth century, Freud built an impressive philosophical system, which he called "psycho-analysis," that revolutionized the human sciences. With its powerful new concept of the unconscious, this system displayed the human psyche in the full splendor of its incontestable abjectness, with all its neuroses and psychoses, with all its drives and animal-like impulses, with its interminable slipups, dead spots, and faux pas. Freud, however, understood that to plumb the innermost depths of the human soul and uncover its most cherished secrets, he would need a series of practical techniques to bypass the many "censors" that, according to him, our own psyches throw in our way. One of these techniques was dream interpretation, which Freud thought could be used to strip away the layers of psychic censorship that distorted the true, or "latent," meaning of our dreams. By subjecting the dreams of his patients to rigorous interpretation, Freud hoped to name their afflictions, alleviate their suffering, and ultimately return them to normal life.[38]

Versions of the pons activation hypothesis existed already in the late nineteenth century. Freud was familiar with them but rejected them all. Having interpreted "more than a thousand dreams" over the course of his medical training, he could not bring himself to accept the idea that the scenes that visit us in the night are meaningless organic events comparable to circulation of the blood or the rumbling of an empty stomach. These scenes have too much power over us to be purely organic. Psychologically, they can build us up or tear us down. As such, Freud reasoned, they must have tremendous emotional and psychological significance. Otherwise, how do we explain the undeniable connection between them and our sense of self? How do we explain the obvious fact that *what* we dream about is tied to *who* we take ourselves to be? Freud concluded that dreams cannot be meaningless organic occurrences, as the organic theories of his day asserted. Quite the contrary, they must be reflections of our psychological and emotional states, echoes of our most private phobias, our most disquieting traumas, and our most clandestine desires. Dreams must be windows into the innermost chambers of the mind that call out for interpretation. They must be, as he memorably put it in *On the Interpretation of Dreams* in 1899, "the royal road to the unconscious."

Freud's clinical revival of the ancient art of dream interpretation had an interesting cultural spillover effect. After 1899, anything related to the study of dreams came to be associated in the popular imagination with Freudian psychoanalysis. One only had to pronounce the word "dream" in intellectual company to evoke an image of a psychoanalyst hunting for the latent meaning of the dreams of a disturbed patient.

Unfortunately, the field of psychology underwent a series of value transformations in the first half of the twentieth century as psychologists sought to bring their discipline in line with the bench practices of the physical sciences. By the 1950s, in fact, the values of the psychological community had shifted so drastically that most psychologists now looked at psychoanalysis with suspicion rather than curiosity, and they discarded anything associated with it, including dream interpretation, as a lousy blend of metaphysical speculation, wishful thinking, and dubious clinical work. As Jacob Conn argues, if by 1939, the year Freud died, it seemed like "the Freudian revolution had achieved all its goals," by the 1950s it seemed as if the revolution had run out of steam.[39]

It is against this broader cultural and historical background that we must interpret the emergence—or, rather, the *re*-emergence—of the pons activation hypothesis in the 1950s. Multiple historians of dream science have pointed out that although this hypothesis was not without empirical support in the nineteenth and twentieth centuries, its appeal in the 1950s was mainly ideological. By reducing dreams to physiological events without psychological significance, this theory allowed empirically minded psychologists who were interested in dreams to kill two birds with one stone. They could defend the scientific status of their endeavors by claiming that dreams were perfectly legitimate candidates for scientific investigation while distancing themselves from psychoanalytic theory, which had now lost its clout. As a result, anyone who dared to speak about the "meaning" of dreams in the 1950s, 1960s, and 1970s was treated as a charlatan, no better than one of those knavish soothsayers who claim to read a person's future in the palm of their hand. Like divination, dream interpretation was nonscientific gibberish masquerading as scientific fact.

Freud's Return

The delegitimization of psychoanalysis through the reduction of dreams to meaningless physiological events would have succeeded had it not been for the pesky fact that dream science took a series of unexpected turns in the 1970s. After years of assurances from dream theorists that dreams could not be interpreted because they were the result of anarchic activity in the pons, scientists discovered that dream formation depends on the activation of key cortical areas as well, especially in the parietal and frontal lobes. The pons activation hypothesis that Freud rejected in his day and age and that was subsequently weaponized against his psychoanalytic project after his death thus became clearly inadequate—or at least incomplete—as an explanation of dreaming. A complete account would have to explain this cortical involvement.

Two discoveries in particular shook the world of dream research. One was the realization that damage to the parietal lobe, the region of the mammalian cortex that helps animals represent and navigate physical spaces, can impede the formation of dreams. This meant that dreams could not possibly be meaningless white noise from the perspective of the dreamer since they involve high-resolution renditions of physical spaces with a clear spatiotemporal organization. The other was the discovery that the brain's so-called "id system," a complex set of neural networks that include the frontal lobe (especially the ventromedial prefrontal cortex [vmPFC]) and the limbic system (particularly the amygdala), is also conscripted into the dream-formation process.[40] Since this system is crucial for emotional regulation, decision-making, and self-control, our dreams must be emotionally colored and thus must have meaning in the context of our personal lives.

These discoveries sent the pons activation hypothesis into a tailspin and triggered what some experts have called the "Freudian renaissance" of the 1980s. Research on the vmPFC and the limbic system was crucial in reintroducing Freud's theories about the emotional weight of dreams into the scientific mainstream, particularly his claim in *On the Interpretation of Dreams* that our dreams are modulated by past emotional experiences. We might say that in the 1980s dream science experienced what Freud calls "the return of the repressed," except that in this case what had been repressed was Freudianism itself. As Mark Solms bluntly puts it, "[it] became clear that neuroscience owed Freud an apology."[41]

Nowadays, few researchers would stand by the theory that dreams are merely organic or physiological events. Most agree that dreams are intimately bound up with emotions. Derek Brereton explains that most of our dreams involve a dream ego immersed "in emotionally salient social space."[42] This is something we have all experienced. When we dream, we dream of places we have seen, people we know, and things we either love or hate. The world of dreams may be strange and unpredictable, but it is far from emotionally neutral. It is circumscribed on all sides by an emotional horizon that makes us relate to it not just as a field for the performance of actions, but also as a field for the cultivation, elaboration, and management of our emotions. Feelings, affects, and sentiments are what dreams are made of, which is why we never just witness our dreams. We embody them. We enjoy or suffer them. We *live* them.

In *The Mind at Night: The New Science of How and Why We Dream*, the science journalist Andrea Rock notes that the influence of the limbic system implies that dreams truly are a "royal road" of sorts—if not to the unconscious as Freud thought, then at least to the emotions. Even against our will, they express what moves us, what drives us, and sometimes even what undoes us:

> Thanks to brain-imaging studies showing the dreaming brain in action, it is also becoming clear that in dreaming consciousness, the limbic system—the command center for directing emotion and storing strong emotional memory—is directing the dream show [. . .] Having the brain's emotional center in the driver's seat means that the memories that are being singled out most prominently for processing [in dreams] are those that are emotionally

charged: anxieties, feelings of loss, blows to self-esteem, and physical or psychological trauma.[43]

In *Integral Dreaming: A Holistic Approach to Dreams,* dream theorists Fariba Bogzaran and Daniel Deslauriers expand upon this idea with two useful metaphors. They say that dreaming is an "emotional change sorter" that stamps past experiences as positive or negative via limbic processes.[44] Because of it, our memories retain their emotional currency. Dreaming, they go on to say, is also a process of "emotional metabolization." Once experiences are stamped as emotionally pleasing or upsetting, dreaming integrates them into our ongoing sense of who we are. Through the act of dreaming, we construct our sense of self: I am someone who fears such and such, who worries about such and such, who yearns for such and such.[45]

The Royal Road to Animal Emotions

If our dreams are the royal road to our emotions, are the dreams of animals the royal road to theirs? The neuroscientist Antonio Damasio gestures in this direction in his book *The Feeling of What Happens*, where he explains that other animals also experience intense emotions during sleep, presumably in the context of dreams. He writes, "Deep sleep is not accompanied by emotional expressions, but in dream sleep, during which consciousness returns in its odd way, emotional expressions are easily detectable in humans and animals."[46]

Contemporary animal sleep research supports this view. Consider an experiment on rat sleep conducted in 2015 by an interdisciplinary team of researchers led by the neuroscientist Freyja Ólafsdóttir from University College London. Like Louie and Wilson (whose research we encountered in chapter 1), Ólafsdóttir and her collaborators exposed a group of rats to a spatial task and compared the patterns of hippocampal activation elicited during waking and sleeping periods. But they controlled for a variable that Louie and Wilson did not: emotional motivation. Are rats more likely to "replay" a spatial task if they are emotionally invested in its solution? Is desire a driving force in the generation of rat dreams? This may sound like the sort of question that no empirical protocol could ever resolve, but Ólafsdóttir and her team came up with a clever two-stage experiment that put the desire of their four-legged research subjects front and center.

In the first stage, they acclimated the rats to a T-shaped track where access to the two smaller arms of the track was blocked by a transparent barrier. Rats could move up and down the base of the maze and see the two branching arms, but they could not physically explore them. The authors then introduced motivation into the picture by cueing one of the arms with a reward (a few grains of rice), while leaving the other one empty. This caught the attention of the rats, who ran up to the junction of the maze and stared longingly at the delicious pile of rice just out of reach. Once familiar with this setup, the rats were removed from the maze and allowed to take a nap. While they slept, researchers recorded what was occurring in their hippocampus, keeping a watchful eye on the order in which various hippocampal cells fired, which resulted in a "neural map" of what the rats were experiencing. But what were the rats experiencing? What was this neural map a map *of*? Ólafsdóttir and her colleagues suspected that the rats were mentally "*pre*-playing" the act of physically exploring the cued arm of the maze and getting their little paws on their object of desire.

To see if their suspicion was right, in the second stage of the experiment they reintroduced the rats to the maze, except that this time the transparent barriers and the pile of rice had all been removed. Upon reintroduction, and as predicted, the rats ran to the junction of the T-shaped maze and immediately turned in the direction of the previously baited arm, suggesting that they remembered which arm held the sweet reward and anticipated finding the reward there. Even after realizing that the rice was not there anymore, the animals spent considerably more time exploring this arm compared to its counterpart.

As the rats shuttled up and down the previously cued arm, the researchers recorded hippocampal spike events and found that the pattern associated with physical exploration of this specific part of the maze was identical to the one they recorded while the rats were napping. The same hippocampal cells fired—and fired *in the same order*—when the rats slept after seeing but not physically exploring the cued arm as when they explored this arm after their nap.[47] This established beyond doubt that the hippocampus was doing the same thing at only two moments: when the rats slept after having seen the reward, and when they explored the space that, much to their disappointment, no longer contained it. The rats, in other words, remembered aspects of their physical environment that piqued their emotional interest and actively imagined a "future experience" in which they their desires were fulfilled. This act of imagination took place while they were fast asleep.[48]

To be fair, it is possible that there is no link in this case between "pre-play" and dreaming since the former took place during slow-wave sleep, when dreams are less likely to occur. But if there is a link, and there are signs that this is the case,[49] there is a lot we could say about these findings. For starters, they seem to vindicate George Romanes's claim from the late 1800s that animal dreams are proof of imagination. While they napped, the rats had to envision what it would be like to traverse a space they had never visited before. For that, they could not just retrieve old memories and replay them; they had to create new subjective experiences using bits and pieces of old ones. Here, imagination must take the reins of the cognitive process and combine old images into news ones, as the French Enlightenment philosopher Voltaire would say, "in endless diversity."[50] I think Ólafsdóttir and her team describe the mental operations of the sleeping rats as acts of "*pre*-play" rather than "replay" because they understand that the rats were envisioning a possible scenario they had never encountered in the real world. The rats were not recalling; they were *projecting*. They were performing acts of what cognitive scientists refer to as "mental time travel," which is the ability to "mentally project oneself [. . .] forward in time to *pre*-live possible events in the future."[51]

Ólafsdóttir's findings also bring emotion into the conversation since this mental time travel did not unfold in an affective vacuum. On the contrary, it was obviously driven by past emotional experiences. The emotional biasing (or "cueing") of the arm is what led rats to dream about the rice in the first place, echoing the contemporary view that dreams and emotions form a double helix structure whose integrity crumbles if we try to untwist its strands. We saw in chapter 1 that rats routinely dream of emotional experiences they have already had, but now we must add that they also dream of emotional experiences they *want* to have. They dream of what their little rodent hearts desire.

Shall we say then that dreams are means of Freudian "wish fulfillment," even for rats? Some may reject this suggestion out of hand, but we cannot deny that, in light of these findings, Freud's remark from chapter 3 of *The Interpretation of Dreams* sounds uncannily modern:

> What animals dream of I do not know. A proverb for which I am indebted to one of my pupils professes to tell us, for it asks the question: "What does the goose dream of?" and answers: "Of maize." The whole theory that the

dream is the fulfillment of a wish is contained in these two sentences.[52]

as does the footnote that supplements it:

> A Hungarian proverb cited by [Sándor] Ferenczi states more explicitly that "the pig dreams of acorns, the goose of maize." A Jewish proverb asks: "Of what does the hen dream?"—"Of millet."[53]

To this, let me add a proverb of my own: "And of what do rats dream? Of rice, of course—at least in human laboratories."

Animal Nightmares, the Horror of Sleep

Sadly, not all the dreams of animals are blissful and uplifting. Some are dark and lacerating. Such is the heartbreaking case of animal nightmares, which put a doleful twist on the early Christian apologist Tertullian's characterization of sleep as "the very mirror of death." Yet, it may be precisely in the nightmares of animals, perhaps more so than in their other dreams, that we observe with the most unnerving clarity the emotional intensity of their inner lives.

In a study published in *Nature* in 2015, a group of Chinese scientists found that rats experience unsettling nightmares after prolonged exposure to physical and psychological trauma. Led by Bin Yu, an expert in neuropharmacology and behavioral neuroscience at Peking University, the researchers put rats in an enclosure and used a transparent barrier to divide them into two groups. The first group was physically tortured with electric shocks to the feet, a deeply sensitive part of the rat body. The second group was psychologically tortured by being forced to watch the first group being tortured on the other side of the transparent barrier. Members of the first group were given electric shocks of increasing intensity every ten minutes. Meanwhile, members of the second group watched helplessly as their friends jumped, struggled, screamed, and eventually uncontrollably urinated and defecated from the pain. The rats were subjected to this grisly combination of physical and psychological violence until they were "modeled" (a term scientists use to describe the process of habituating animals to a stimulus), at which point they were finally removed from their torture chamber.[54]

The experiment, a nightmare of its own, did not end there. Twenty-one days later, the rats were reintroduced into the same enclosure to see whether they remembered the site of their trauma. As soon as they stepped inside, they displayed what the authors describe as "total freezing behavior." They did not move, walk, pace, or run; they did not scream, bite, or play dead; they did not cower in the corner or attempt to flee. They froze. Everything about them—"with the exception of respiration," we are told—became like a statue.[55]

Then came the nightmares.

When they fell asleep, the rats had such terrifying dreams that they regularly woke up in a panic, a behavior that the researchers dubbed "startled awakening." EEG analysis of brain activity in the moments leading up to startled awakening demonstrated that the rats were experiencing nightmares triggered by traumatic memories. It seems that at some point in their sleep cycle, they would retrieve memories of their not-too-distant past and re-live them in the form of a dream. Since the memories were emotionally painful, their retrieval activated their amygdala, arousing a sharp feeling of fear. In fact, the memories were so emotionally damaging that they disrupted the neural circuits that normally keep the amygdala within normal range, especially the infralimbic and ventral anterior cingulate cortices, resulting in a "dis-inhibited" or "hyper-functioning" amygdala. As a result of this disinhibition, the rats did not just experience fear. They experienced an *accumulation* of fear—a fear that grew impatiently, with no sign of relenting.[56]

Under normal conditions, rats would respond to this accumulation of fear by activating their "fight or flight" system. Unfortunately, trauma wrecks this system as well, leaving animals trapped in state of high alert, unable to respond to their environment via fight or flight. In *The Body Keeps the Score: Brain, Mind, and Body in the Healing of Trauma,* the Dutch psychiatrist Bessel Van der Kolk explains that when an organism faces a threat to its wellbeing, it has three general responses at its disposal. The first is "social engagement," which entails reaching out to others for help. When this option fails, the organism can engage the "fight or flight" response, especially if the threat is serious and imminent. Unfortunately, there are times when the situation is so dire that the organism can neither fight nor flee. In these cases, the organism has no choice but to activate "the ultimate emergency mechanism," at which point it "disengages, collapses, and freezes,"[57] effectively shutting itself off in a last bid to stay alive.

This, I believe, is what happened to the rats in Yu's experiment. When they were reminded of their trauma by being reintroduced to their enclosure, their bodies came to an involuntary, sudden, and brutal halt. And when they were reminded of their trauma in their sleep through mental replay, they entered what their human handlers described as "an emergency situation"—a situation so catastrophic that the rats had no choice but to go into a state of shock and convulse themselves awake.[58] One of the most depressing aspects of this whole affair is that even if these poor animals eventually woke up from their nightmares, they never "woke up" from their trauma. Once they were "modeled," the horizon of their lives shriveled to nothing. From that moment onward, they could only do one thing: spend the rest of their days alternating between waking and dreaming reenactments of their own brutalization.[59]

Aside from revealing something frightening about our collective willingness to mutilate animals in the name of science, these nightmarish experiments teach us that trauma disfigures the emotional profiles of living beings to the point that they start mimicking the behavioral symptoms of post-traumatic stress disorder (PTSD), which include constant recall of upsetting memories, sudden flashbacks, emotional distress after exposure to triggering reminders, difficulties sleeping, heightened startle reaction during sleep, and chronic nightmares.[60] Nightmares in particular can emotionally scar animals and hinder their cognitive functioning. Because they fortify excruciating memories through constant replay, nightmares make animals more likely to develop generally maladaptive sleep patterns that cause them to lose their focus and sink into freezing behaviors while awake.[61] As the Canadian psychiatrist Laurence Kirmayer observes, the relationship between trauma and nightmares is bidirectional, "with trauma evoking nightmares and nightmares leading to increased thoughts about trauma."[62]

It is not just rodents that can be caught in this fatal psychic loop. Elephants can suffer a similar fate. Because of their formidable memories and complex social lives, young calves who experience intensely traumatic events, such as witnessing poachers butchering their mothers and cutting out their tusks with an electric saw, store these horrific images in their long-term memory. Later, they dredge up these memories against their will, developing what can only be described as PTSD.[63] These memories "haunt them during the day [in the form of flashbacks] and often return at night in the form of nightmares and night terrors, *re*-traumatizing the calves."[64]

Nightmares destroy the emotional stability of young elephants by locking them into a rotten cycle in which they get no reprieve from the original traumatic event. Jeffrey Masson reports the effects of such trauma in his book *When Elephants Weep: The Emotional Lives of Animals*:

> Animal behaviorists are unlikely to acknowledge that terror can return in the dreams of animals. And yet from a Kenyan "elephant orphanage" comes a report of baby African elephants who have seen their families killed by poachers and witnessed the tusks being cut off the bodies. These young animals wake up screaming in the night. What else but the nightmare memories of a deep trauma could occasion these night terrors?[65]

Barbara King, a biological anthropologist who specializes in animal emotions, describes the life of one of these orphans named Ndume, who ended up at the elephant nursery in Kenya's Nairobi National Park:

> [Ndume] was a baby elephant living wild with his family in Kenya. When the family wandered from the forest into an area seeded with crops, the elephants were attacked, and many were killed by angry farmers wielding spears and arrows. Ndume himself managed to flee. However, he witnessed a smaller calf near him hacked into pieces and suffered from shock and from the knife gashes he himself experienced. Ndume was brought to an elephant sanctuary outside Nairobi called the David Sheldrick Wildlife Trust. Three months of age at the time of the attack, he began to cry and bellow for his dead mother after his arrival at the Trust. He could not sleep well. Sanctuary experts believe he was reliving the trauma of the attack in his dreams. Then Ndume became depressed.[66]

FIGURE 8. Orphaned calves at an elephant nursery in Nairobi, Kenya, are woken up by nightmares and roam the grounds at night searching for their mothers, who were killed for their tusks by ivory traffickers. Many of these calves become severely depressed.

At night, Ndume was so apprehensive and inconsolable that he would bellow as loudly as he could until his keepers let him out of his sleeping quarters, at which point he would rummage around in the dark, frantically looking for a mother who could never be found.[67] Parallel reports of elephant nightmares have come out of other sanctuaries, such as the elephant rehabilitation center at South Luangwa National Park in Zambia.[68]

The same goes for nonhuman primates. The American psychologist Francine "Penny" Patterson, who became famous in the 1970s and 1980s for teaching gorillas American Sign Language (ASL) at Stanford University, recounts that one of the gorillas under her care, Michael, was ravaged by nightmares of early childhood trauma. He often woke up screaming in the middle of the night, sometimes signing to Patterson immediately afterwards, "Bad people kill gorillas."[69] Like the calves in Kenya and Zambia, Michael witnessed the murder of his mother at a young age, this time at the hands of the same Cameroonian bushmeat traffickers who later sold him into captivity when he was no older than three. When asked, as an adult, "What can you tell us about your mother?" Michael responded with the following sequence of ASL signs:

> Squash, gorilla mouth, tooth
> cry
> sharp noise, loud
> bad
> think-trouble
> look-face, cut, neck, lip, girl, hole.[70]

For as long as Patterson remembers, Michael has been terrified of men "working in and cutting trees."[71]

The animal studies scholar Concepción Cortés Zulueta interprets Michael's string of signs as a nonhuman "enunciation of trauma," which is consistent with evidence from comparative psychology that primates who lose their mother at a young age develop physiological, behavioral, and psychological disturbances that follow them for the rest of their lives. Maria Botero, a philosopher and psychologist who specializes in mother-infant relations across species, has argued that the loss of a mother can be world-shattering for young primates:

> The lack of a mother can have multiple behavioral and neurophysiological effects throughout the orphan's life. In several species, orphans can experience decrements in growth, reproduction, and longevity and suffer from negative impacts on health, social status, and emotional development such as anxious behavior and effects on their ability to engage in successful social interactions, with declines in play and other social responses and increases in abnormal behavior, such as lethargy, rocking, and hair plucking.[72]

These effects are particularly pronounced for primates who, like Michael, "become orphans as the result of bushmeat practices and the pet trade,"[73] and even worse for primates who, also like Michael, are subsequently isolated from other conspecifics and kept in captivity for prolonged periods of time.[74]

That nighttime can be grim and full of terrors for animals raises profound questions about their emotional lives. Any animal who succumbs to frightening visions when the sun goes down must be capable of storing episodes from their past in long-term memory, dredging up these episodes at later times, and experiencing potent emotions like fear, aggression, panic, anxiety, and terror in relation to them.[75] These emotions are living testaments to the social attachments these animals need to flourish. Above all else, they bring into focus the affective scaffolding that organizes their lives and gives meaning and structure to their experience of the world.

It is hard not to be overpowered by the facts here. We have heard about rats afflicted by the horrors that human scientists have imprinted onto their brains; about calves who wake up screaming in the middle of the night facing the gray prospect of a life without a mother; about a gorilla so wounded by human greed that he apparently transferred his primal fear of Cameroonian bushmeat traffickers to university staff in Palo Alto, California. We have heard, in short, about the most devastating kind of emotional damage, the kind that stalks and hunts down its prey until they perish in its maw. And all in the context of dreams.

METACOGNITIVE CONSCIOUSNESS: CAN ANIMALS LUCID DREAM?

When we dream, we typically experience the events unfolding before us as slices of real life, even when they run afoul of the most basic laws of logic and physics. This is because dreaming curtails our ability to reflect upon our own experience, leaving us unaware of the fact that we are dreaming. This lowering of our metacognitive guards is such a prominent feature of dream phenomenology that it has led most philosophers since the seventeenth century to interpret dreams first and foremost as epistemological problems—as obstacles we must overcome on our path to genuine knowledge. As Descartes wondered in *Meditations on First Philosophy* in 1641, how can we know anything with certainty when we have no way of knowing whether we are awake or trapped in a dream? How can we trust our senses when they seem to conspire against us, doing everything in their power to keep us from telling the difference between the true and the false, between dream and reality?

While this facet of dreaming merits further analysis (especially since it raises provocative questions about the putative centrality of rationality in human life), when we dream, we are not *always* in the state of metacognitive impairment that vexed the father of French modernity. Occasionally, we regain our metacognitive faculties in the middle of a dream sequence and, in a sudden flash of mental clarity, become aware of the fact that we are dreaming. As one of Descartes's philosophical contemporaries, the German philosopher Gottfried Wilhelm Leibniz, observed in his *Catholic Demonstrations* from 1668: "Now and then, the dreamer himself observes that he is dreaming, yet the dream continues. Here he must be thought of as if he were awake for a brief interval of time, and then, once more oppressed by sleep, returned to the previous state."[76] When this happens, the dreamer experiences a "lucid dream,"[77] a peculiar kind of dream whose most salient phenomenological property is the presence of "metacognitive insight into the fact that one is now dreaming."[78] Far from impairing our metacognitive abilities, lucid dreams give us a heightened sense of mental dexterity, wonder, and even freedom. In them, we even see "with startling clarity that what seemed an unquestionably external, objective, material, and independent world is in fact an internal, subjective, immaterial and dependent creation of mind."[79] It is no wonder that so many dream experts wax poetic when talking about lucid dreams and describe them as "magical" and "miraculous"—because these dreams propel us into a mesmerizing plane of existence in which our minds somehow manage to pull back the dream's veil of illusion while, as Leibniz says, "the dream continues." On this surreal plane, we are physiologically asleep yet cognitively awake.

In this section, I round off my presentation of the SAM model of consciousness by considering whether the experience of dream lucidity, which is frequently seen as an instance of metacognition, is inimically human or whether it, too, could lie within reach of other animals' minds.

Lucid Dreaming: The Exception to the Rule

In cognitive science and philosophy of mind, lucid dreams are widely accepted as instances of "metacognition" (from the Greek *meta*, meaning "from above"), which denotes a unique form of awareness that involves thinking about thinking or being aware of being aware.[80] In an influential article on the subject, the psychologist Tracey Kahan explains:

> Dreams in which the dreamer becomes aware of dreaming while continuing to dream are known as "lucid dreams" [. . .] Attaining lucidity in dreams requires evaluation of experiences as they happen in the dream, a process termed "metacognitive monitoring." Metacognition includes, but is not limited to, the monitoring of one's thought processes and the deliberate direction of them.[81]

In Kahan's interpretation, lucid dreams are metacognitive undertakings because, in them, the dreamer redirects their intentionality (that is, their mental focus) from the contents of their mental state to their mental state as a whole. In other words, the dreamer stops attending to the things that appear in their dreams and starts attending to the mode in which these things appear—namely, as dreams.

The scientific consensus appears to be that only humans can experience dream lucidity because only we are metacognitive agents capable of thinking about our own thoughts and becoming aware of our own awareness. Compared to us, all other animals are prisoners of their own restricted minds, beings condemned to live their entire lives "inside" their mental states, incapable of looking at them "from above." Even though this consensus is starting to break down as more and more evidence of animal metacognition surfaces in the behavioral and cognitive animal sciences, nobody involved in this research has thought to look to the dreams of animals for insight into the nature of animal metacognition.

The reason is simple. For a long time, it has been thought that lucid dreaming requires mastery of those so-called "higher" cognitive capacities that scientists and philosophers have historically praised humans for possessing, especially language, conceptuality, and rationality. Who would ever expect a lowly animal to reach such daunting heights? As the dream experts Ursula Voss and Allan Hobson explain:

> We have no animal model for dream lucidity because we have every reason to suppose that reflective insight such as observed in lucid dreaming necessitates sufficient language capacities assumed essential in the formation of abstract thought or reporting of such. For this reason, we assume that infra-human mammals, which lack significant language capability, cannot become lucid.[82]

Notice that Voss and Hobson's position, which is representative of the mainstream, hinges on a thesis about language. Language hoists us into a realm of mental abstraction where we can play with abstract concepts, which is a precondition for the "reflective insight" characteristic of lucid dreams. Bereft of language, animals cannot access this realm of abstraction and, as such, "cannot become lucid."[83]

I do not deny that we do not currently have an animal model for dream lucidity since I cannot imagine how lucidity in other species would be studied under controlled laboratory conditions, but the point here is philosophical. Why should we accept Voss and Hobson's linguistic interpretation of the mental processes that prop up lucidity? Why couldn't lucidity manifest itself outside a linguistic schema, in the absence of abstraction, conceptuality, and rationality? Who gets to determine whether these capacities are the lifeblood of lucidity or whether lucidity can exist independently of them? And on what basis is this determination made?

In what follows, I argue that there are other ways of thinking about lucidity that are more fruitful from the standpoint of an interspecies theory of dreaming and that open this capacity to other animals. These other ways of thinking draw a contrast of a rough and ready sort between *conceptual* and *pre-conceptual* experiences of dream lucidity, or, in other words, between the experience of lucidity itself and the boundless cognitive operations to which it can, but need not, be subjected.

A Janus-Faced Theory of Lucidity: A-Lucidity vs C-Lucidity

The philosophers Jennifer Windt and Thomas Metzinger make this distinction between conceptual and pre-conceptual experiences of lucidity and criticize contemporary theories of dreaming for blurring the line between them. They call these experiences:

A-lucidity ("A" for "attention"), which occurs when the dreamer experiences "spontaneous insight" into the fact that they are dreaming but without performing any additional cognitive or metacognitive operations; and

C-lucidity ("C" for "cognition"), which happens when the dreamer experiences "spontaneous insight" into the fact that they are dreaming and then uses this insight as a springboard for additional cognitive or metacognitive operations, such as conceptual judgments, rational inferences, or linguistic reports.[84]

Though closely related, these types of lucidity are distinct. In the first, the dreamer simply notices that they are dreaming, while in the second they notice this *and* use this insight as fodder for more clearly cognitive operations, such as exercising voluntary control over what happens in the dream world, making inferences about their own mental states, or even forming conscious mental judgments of the form "This, here, is a dream." In A-lucidity, the dreamer attends; in C-lucidity, they attend and cognize.

Windt and Metzinger see the relationship between these two types of lucidity as one of unidirectional entailment, meaning that all instances of C-lucidity necessarily entail A-lucidity but not the other way around. In their interpretation, where other researchers go astray is in assuming that just because the lucid dreams they happen to study in their laboratories—the dreams of neurotypical, adult humans—usually incorporate attentional and cognitive components, that both of these components must be found in *all* lucid dreams. Unfortunately, this assumption commits what philosophers call "the mereological fallacy," which is an error in reasoning that stems from attributing properties to the whole that apply only to the part. This fallacy leads dream researchers to believe that what is true of *some* lucid dreams is true of *all* of them, and thus to see the relationship between A- and C-lucidity as one of bidirectional entailment, which is a mistake. As Windt and Metzinger say, in a lucid dream one cannot cognize without attending, but one *can* attend without cognizing. That's what A-lucidity is. It is lucidity without cognition. It is dream meta-consciousness without language, concepts, or reason.

The connection to animals here is tricky, so allow me to begin by noting that Windt and Metzinger explicitly say that C-lucidity applies only to "rational creatures capable of self-directed concept formation," which in their eyes includes only humans. Here, they follow the mainstream view. I suspect some animal advocates might see this conclusion as rushed and wonder what Windt and Metzinger might mean by "concept" and "rational" given that there is a bevy of evidence in comparative psychology that plenty of other animals form abstract concepts and perform rational operations, including logical inferences and mathematical calculations.[85] From the standpoint of this evidence, it is not clear that only humans form self-directed concepts (although, of course, this depends on how these terms are defined). While this line of thinking is a valuable reminder that we must not judge animals prematurely, I believe there is another angle worth pursuing. What if, instead of worrying about the exclusion of animals from this more sophisticated type of lucidity (C-lucidity), we focused on their inclusion in its more lenient counterpart (A-lucidity)? What would it mean for us to accept that animals experience A-lucidity in their dreams? What would this tell us about their powers of reflection, and about their status as metacognitive agents more generally?

Windt and Metzinger define A-lucidity in two ways. At one point, they define it as "spontaneous insight," observing that a dreamer has an A-lucid dream when they suddenly realize that they are dreaming, even if this realization does not lead to cognitive or metacognitive operations down the road. At another, they define it as "introspective attention." In an A-lucid dream, the mind of the dreamer folds upon itself, becoming subject and object simultaneously. In this pleated state, the mind introspects. It looks inside itself. With these two images, Windt and Metzinger paint a picture of A-lucidity as a mental process whereby the dreamer pays attention to their mental states while dreaming, thereby transforming these states into the content of a new, higher-order mental state.

By most standards, there is nothing strange about this account of A-lucidity. What is strange is Windt and Metzinger's assertion that this "is something many animals could possess."[86] In their eyes, apparently, many other species could turn their mental focus inward and pay attention to the dreamlike quality of their experience during sleep. These species could even undergo some sort of "insight" about this experience, even if they never perform any of the cognitive gymnastics that human dreamers do, such as making rational inferences, playing with abstract concepts, or forming complex mental judgments. Unfortunately, Windt and Metzinger never tell us which animals might have this extraordinary mental power since they make this comment in passing, without teasing out its implications. But we should not allow this lack of specificity to distract us from the crucial point here, which is that two influential figures in contemporary dream research believe it is perfectly possible for other species to experience mental clarity in the middle of a dream sequence, the same sort of clarity that Kahan would file under the category of "metacognitive monitoring." Even without all the details worked out, this is a truly radical concession from the standpoint of the philosophy of animal consciousness, for it means that other animals may be, like us, metacognitive agents who become aware of their own awareness—and in their dreams to boot!

Animal Metacognition: From Conceptual Judgment to Embodied Feeling

Most dream experts are convinced that only humans can have lucid dreams because dream lucidity is defined as always involving two moments:

1. a *moment of dissociation* when the dreamer takes a step back from the perceptual field in which they are immersed to monitor the field as a whole, which amounts to a re-orientation of the dreamer's intentionality;
2. a *moment of judgment* when the dreamer subsumes a concrete particular under an abstract concept, culminating in a mental judgment (e.g., "This is a dream").

While many find this interpretation appealing, I worry that it relies on philosophically loaded concepts that obfuscate more than they clarify, such as the concept of "judgment." In the eyes of many academic philosophers today, the term "judgment" excludes animals *a priori* as it denotes a subjective attitude with a propositional structure (in other words, subject plus predicate).[87] Since animals do not make propositions with subjects and predicates, they are thought to be inherently incapable of forming mental judgments. Even without knowing much about animal sleep research or dream science, these philosophers categorically deny that animals can experience lucid dreams. But we must ask: where does this denial come from? Does it come from a careful study of the phenomenon under consideration, or from an uncritical acceptance of a loaded philosophical concept? What is doing the work here: the theory or its terminology?

To appreciate the gravity, consider what happens when we replace, in the second moment above, the term "judgment" with the term "feeling." We end up with more than a new description of lucidity; we end up with a new conception of it—something that quickly begins to resemble Windt and Metzinger's concept of A-lucidity. Under this revised version, animals could be said to have a lucid dream if they experience:

1. a *moment of dissociation* during which they take a step out of the perceptual field in which they are immersed to attend the field itself; and
2. a *moment of feeling* during which they sense, in an affective and embodied manner, that this field is somehow different from the field they perceive when awake.

Once this affective and embodied feeling replaces the cognitive act of judgment, it becomes easier to see how animals might be able to experience lucidity in their dreams, even if their experience of the world is not mediated by language, conceptuality, or rationality. Here lucidity appears as a pre-conceptual and pre-cognitive feeling that spontaneously takes over the mental life of the animal during a dream.[88]

What does this feeling look like? It is impossible to say for sure, but allow me to offer a plausible scenario. It could be that animals are familiar enough with their field of waking perception to tell, intuitively, when something about this field is amiss, as happens in dreams. Perhaps its contents are sufficiently bizarre to raise a flag. Perhaps its inner organization is sufficiently strange to trigger a different response.[89] Either way, this feeling of perceptual incongruity might prompt animals to pay attention to the dream state rather than to its contents, bringing about a moment of dissociation. If animals realize on a pre-conceptual level that this perceptual field is different from their wakeful one, this realization would amount to a "spontaneous insight." Perhaps this is how lucidity manifests in animals: less as a conceptually laden act of judgment and more as a gut feeling that something about their phenomenal experience is amiss, that something simply does not add up. As the philosopher Mark Rowlands explains, many animals do not need complex mental judgments to know that "something is up."[90]

To be clear, I am not saying animals experience lucid dreams *as* lucid dreams, which is the ambit of C-lucidity. I am not even saying they experience A-lucidity as a matter of fact. I recognize that the idea of animal lucidity is speculative and that, when we invoke it, we leave solid ground. Nevertheless, this idea may be less far-fetched than many suppose, and for a few reasons. First, as we have seen, there are scientifically grounded theories of dreaming that accommodate it.[91] Second, the rapidly growing body of literature on animal metacognition is teaching us that many species show signs of being aware of being aware.[92] The question, then, is no longer whether meta-consciousness is possible for animals, but whether it is possible, as Michel Jouvet says, "in the labyrinth of sleep." Third, research in functional neuroanatomy proves that many animals, especially mammals, have brain structures that are evolutionarily homologous or functionally analogous to those that bring about lucid dreaming in humans, especially the dorsolateral prefrontal cortex.[93]

To this, we should add another observation: even those who entertain doubts about animal lucidity are not always consistent on this point. In the passage quoted above, for example, Hobson and Voss are steadfast on their belief that there is no "animal model" of lucidity because animals lack language, which is a precondition for abstract thought. Yet, in an earlier publication they struck a remarkably different tone, arguing that some animals, especially birds and primates, could indeed experience "awareness of awareness" during sleep.[94] I do not know why they changed their minds, but if their earlier assessment turns out to be right, it could dramatically change our perception of the minds of animals. For it would mean that there could be creatures out there—perhaps the crows that fly over the grasslands of the American southwest, the chimpanzees who rove the forests north of the Congo River, or the black kites who use fire to flush out prey from the fields of Australia's Northern Territory—who, like us, wake up *to*, not just *from*, their dreams; creatures who also pull back the dream's veil of illusion even as, to quote Leibniz one final time, "the dream continues."

ADVANTAGES OF THE DREAM APPROACH

In this chapter we have taken several steps toward a theory of animal consciousness using dreams as our lodestar. By attending to the subjective, affective, and metacognitive dynamics of what Paul Manger and Jerome Siegel dub "mentation during sleep,"[95] we have sought to deepen our understanding of the minds of animals while at the same time stirring contemporary theories of animal consciousness from their dogmatic slumbers. Now, by way of conclusion, I would like to consider two advantages of this "oneiric" approach.

The first is that it avoids one of the most common objections to animal consciousness, which I call "the behaviorist reduction." This objection holds that we cannot use the behaviors of animals as evidence of conscious awareness because those behaviors might be nothing more than unconscious reactions to external stimuli—reactions rooted in innate reflexes, evolutionary instincts, or learned associations. For instance, I may believe that my dog Osa wagged her tail every time I came home because she was happy to see me, but her tail-wagging could have been a rigid and predictable reaction triggered by something in her environment, evolutionary lineage, or personal past. Maybe that is how all dogs react to other animals after prolonged periods of isolation. Maybe dogs have evolved to behave this way around humans. Maybe Osa learned to associate my presence with a reward, such as being petted, being fed, or being taken out for a walk. Either way, to what extent am I justified in saying she was happy to see me just because she wagged her tail when I walked through the door?

Notice that under each of these explanations, Osa is reduced to a marionette that the puppeteering forces of the external world pull this way and that. She does not act; she is acted upon. She does not think; she processes stimuli. She does not feel positive or negative emotions; she responds to rewards and punishments. What makes this reduction dangerous, in my view, is that it is always possible to interpret the behaviors of animals *behavioristically*; that is, as detached from any mental representations or internal phenomenology. Depending upon one's background assumptions and theoretical commitments, one can always boil even the most complex behaviors of a living organism down to simple and predictable reactions, especially if one disregards what is lost along the way.

Dreaming, however, throws a wrench into the conceptual machinery of this reduction. No one, I suspect, would argue with sincerity that a dog dreaming of chasing a ball or a cat dreaming of fighting an enemy are reacting to the mandates of external nature since there is no external nature to react to in this case. Here, the ball and the enemy are endogenous phenomena that animals conjure up by sheer mental power. They are figments of these creatures' imaginations. As such, they call for an explanation that exceeds what the behaviorist reduction can offer. As the cognitive neuroscientists Martina Pantani, Angela Tagini, and Antonino Raffone have argued in connection to human dreams, dreams are imagined contents that pose a formidable challenge to those theories of consciousness that strive to do away with internal mental states altogether.[96]

A second benefit of the oneiric approach to animal consciousness is that it cues us in to the mental freedom we share with so many other species. The philosopher Michel

Foucault, who maintained a keen interest in the history of dream interpretation throughout his life, argued that our dreams bring to the fore our most original freedom, which is our freedom *to transcend*.[97] Dreams catapult us from the realm of the immediate to that of the mediated, from the domain of immanence to that of transcendence. The dream, Foucault said, "[is] the experience of transcendence under the sign of the imaginary."[98] It is "the freedom of man [*sic*] in its most original form."[99]

Admittedly, Foucault did not have the dreams of animals in mind when he wrote this, and we would be hard-pressed to extract from his works anything resembling a theory of animal transcendence. But it is not hard to see how his line of thinking might be conscripted into a non-anthropocentric theory of dreaming. Indeed, we find a more inclusive and animal-friendly version of Foucault's argument in the writings of his compatriot, the philosopher and neurologist Boris Cyrulnik. In a 2013 interview with the French magazine *Le Coq-Héron*, Cyrulnik follows Foucault in claiming that dreams are "escapes from reality" through which living beings liberate themselves from the here and now. "A living being who dreams escapes from immediacy," he says.[100] Unlike Foucault, however, Cyrulnik believes that the transcendence of the dream does not depend on the humanity of the dreamer but on the act of dreaming itself. We do not transcend because we are human; we transcend because we are dreamers. And so are other animals with complex nervous systems, such as cats, dogs, and giraffes. They, too, dream their way to freedom. They, too, produce an entire world-analogue during sleep through something akin to what Sartre calls "a flow of creative will,"[101] a flow that culminates in the negation of *what is* and the affirmation of what *could be*. They, too, in short, experience transcendence "under the sign of the imaginary."

It is hard to deny that dreams push us to new frontiers in the study of animal consciousness, frontiers where the lines between the human and the nonhuman, the existential and the biological, the transcendent and the immanent begin to blur. In this chapter, we have looked at three of these frontiers: subjectivity, affect, and metacognition. In the next, we look at a fourth that has been hovering silently over them for some time: imagination. For, as the Roman philosopher and poet Lucretius says, to dream is to behold in one's mind imaginary shapes of one's own making, wonderous idols "dancing forward in rhythmic measure."[102]

CHAPTER 3

A Zoology of the Imagination

> What this strong music in the soul may be!
> What, and wherein it doth exist,
> This light, this glory, this fair luminous mist,
> This beautiful and beauty-making power.
>
> —SAMUEL COLERIDGE[1]

THE SPECTRUM OF IMAGINATION

Up until now we have focused on dreams, but dreams do not exist in isolation from the rest of an organism's mental life. Indeed, dreams are part of what the philosopher Nigel Thomas calls "the multidimensional spectrum of mental imagination,"[2] a wide band of conscious activity that includes acts of imagination, daydreams, hallucinations, vivid memories, flashbacks, hypnagogic and hypnopompic images that arise just before sleep and awakening, hypnotic visions, playful pretense, and, on some accounts, even waking perceptions. These ebullitions of the psyche may have different neural bases and phenomenological profiles, but they have this in common: they are all children of imagination. It is impossible to discuss any of them, including dreams, without at the same time discussing the larger spectrum of imagination in which they find their home and their truth. As Foucault puts it, even the most thwarted, most elementary dream "opens up onto a new horizon," which is the horizon of *the imaginary*.[3]

Unfortunately, philosophers and scientists have traditionally failed to see signs of the imaginary in other animals, the horizon of their own imaginations being perhaps too narrow to accommodate the imagination of other critters.[4] Even Foucault, whose writings have been used by animal advocates in the global struggle for animal liberation, endorsed a brazenly anthropocentric theory of imagination. Convinced that this faculty is the mainspring of human existence, he went so far as to urge philosophy to transform itself into an "anthropology of the imagination."[5]

Given what we know now about the dreams of animals, however, we can no longer put our faith in an anthropological theory that would cloister the imagination within the bounds of the human. What we need is a zoological theory of imagination that is willing to follow the movements of this faculty beyond the human world, tracing its roots to the very soil of animal life. In this chapter, I make inroads into such a theory with two case studies that reveal important parallels between dreams and other imaginative acts, especially hallucinations, playful pretense, and daydreams. These case studies, one about primates and one about rodents, bring us to the realization that humans are not the only creatures who wend their way through life, as Coleridge describes imagination, "with music in the soul."

MONKEY SEE, MONKEY DO (CASE STUDY 1)

In 1966, the psychologist Gay Luce was hired by the National Institute of Mental Health to write a report for the United States Department of Health, Education, and Welfare (DHEW) entitled "Current Research on Sleep and Dreams." This report was meant to be an up-to-date survey of the major trends in the fields of sleep and dream research, its principal goals being prevention of unnecessary duplication of research and stimulation of cooperation across disciplines, especially psychology, psychiatry, physiology, and anthropology.

A three-time recipient of the American Psychological Association award for journalism, Luce draws readers in from the beginning with her commanding prose:

> Between the darkness out of which we are born and the darkness in which we end, there is a tide of darkness that ebbs and flows each day of our lives to which we irresistibly submit. A third of life is spent in sleep, that most unusual yet profoundly mysterious realm of consciousness where the person seems to live apart from the waking world, often immobile, appearing to be departed. Why all of this sleep? Why does the animal plunge into these periods of stillness?[6]

This is followed by a more or less complete picture of how "astronomers of the mind"—especially psychologists, biologists, and psychiatrists—understood sleeping and dreaming at the time, a picture replete with explanations of the physiology of sleep, the structure of the mammalian sleep cycle, the effects of sleep deprivation, the nature of sleep-related disorders, and the possible origins, causes, and functions of dreams.

In Search of Hallucinations

About eighty pages into the report, there is a section entitled "Animal Dreaming" in which Luce points out that we may not be the only species who frequent this "most unusual yet profoundly mysterious realm." There, she describes what could very well be the first empirical confirmation that animals experience reenactments of waking life while asleep: an experiment on monkey hallucinations conducted in the early 1960s at the University of Pittsburgh by Charles J. Vaughan.[7]

Vaughan put a group of rhesus monkeys into a sensory deprivation booth one at a time and trained them to press a bar with their thumbs at a specific speed whenever an image was projected onto a screen mounted inside the booth—otherwise, they received an electric shock to the feet. Once the monkeys learned to press the bar "at criterion" when presented with a visual image, they were subjected to periods of sensory monotony (lasting about seventy-four to ninety-six hours each), during which they could not hear or see anything at all. Visual stimulation was blocked through the implantation of plastic cornea lenses, while ambient noise was drowned out using a white noise machine. The idea was to block two of the monkeys' most important senses—hearing and vision—to see whether the behavior associated with the presentation of a visual image would suddenly return. Vaughan's reasoning was that if the monkeys pressed the bar at criterion during a period of sensory deprivation, this would mean that they were experiencing a visual hallucination in the dark and, as Luce says, "seeing things."[8]

By the end of his experiment, Vaughan found zero evidence that rhesus monkeys hallucinate while awake; instead, he found something arguably more provocative: that they "hallucinate" while asleep.[9] Naturally, there were times during the periods of sensory monotony when the monkeys fell asleep, and that is when many of them were observed pressing the bar at criterion, suggesting that some kind of visual experience was triggering the conditioned reflex. This behavior occurred during REM sleep. Luce explains:

> During these rapid-eye-movement periods, the monkeys suddenly began pressing the bar at a frenetic pace, rapidly and regularly as in waking. Sometimes they also made facial grimaces, flared nostrils, breathed deeply, and even barked as they pressed the bar. Presumably they were "seeing things" during these intervals of rapid eye movements and were avoiding the shock associated with images. After the isolation period was over, the monkeys

> were tested in the training situation to be sure that they still responded reliably to the projected images. The investigators had seen only one instance of bar pressing during waking, so they had little data about hallucinations, but the evidence that monkeys experience visual images in sleep was very strong.[10]

Luce interprets this as "very strong" evidence that the monkeys were "experiencing some visual reenactment of [their lives],"[11] and concludes by lamenting how little we know "about the interior fabric of consciousness in animals." Convinced that more research of this sort would give us a firmer grasp on this interior fabric, she muses: "Perhaps a next step—training monkeys to respond to particular images or to smells, for instance—may allow us to find out what a monkey dreams about."[12]

I believe that Vaughan's experiment and Luce's description of it raise interesting questions about mental representation in other species. How did the monkeys manage to mentally represent visual images to themselves in the absence of sensory stimulation? And given that many experts in cognitive science and philosophy of mind believe that propositional language is a precondition for having any sort of mental representation at all, how did they manage to arrive at such mental representations in the absence of language? Is it that these animals never generated any genuine mental representations? Or could it be that these experts have been misled—misled by their own human-centric worldview, that

FIGURE 9. Charles J. Vaughan's sensory deprivation experiment in the 1960s accidentally proved that rhesus monkeys experience endogenous visual scenes during sleep. While this research raises important questions about the phenomenology of animal sleep, it also raises ethical questions about the use of animals in psychological research.

is—to believe that there can be no mental representation without language?

Since I do not believe that language is necessary for a rich inner life, I favor this second explanation. In my view, the question we should be asking is not, "Can thought materialize outside the frame of language?" but, "What other frames are there, beside language, for the actualization of thought?" The philosopher Dieter Lohmar, an expert on phenomenology who is critical of linguistic approaches to consciousness, offers an intriguing answer.

Non-Linguistic Representation in Primates

In a 2007 article, Lohmar contends that it is time to move past the old-fashioned notion that the only way to represent the world is through the medium of language, a notion born out of a narcissistic fetishization of human abilities. Surely, language is one medium that animals can use to represent the world, but it is not the only one. Even humans represent their environment "poly-modally,"[13] which is to say, using several representational modes concurrently, including: (1) a linguistic-conceptual mode that employs words, concepts, and propositions; (2) a scenic mode that generates visual scenes with the aid of the imagination (scenes that include people, objects, and events); (3) a gestural mode that operates through the production and interpretation of gesticulations, bodily signs, and facial expressions; and (4) an emotive mode that recalls past emotions, feelings, and bodily sensations. Humans, Lohmar argues, think through *all* these modalities.

In describing mental representation as multimodal, Lohmar opens the door to the possibility of thought without language since it is possible for humans to think about the world nonlinguistically by activating any of the other modes of representation at their disposal, either on their own or in combination with one another. For example, if I activate the visual mode, I can generate endless visual scenes, such as a scene of a room or a landscape. If I activate this mode in conjunction with the emotive one, I can generate visual scenes that evoke strong emotional responses and bodily feelings, such as the scene of a crime or a place that brings me inner peace. Alternatively, by also activating the gestural mode, I can generate visual scenes featuring other people communicating with me by means of signs, grunts, and gestures, such as someone winking at me, shaking their head, or pointing something out to me with their index finger. Even if all these scenes lack linguistic content and linguistic structure, Lohmar says, they remain mental representations of sorts. They remain thoughts that have meaning for me insofar as they represent a certain physical, emotional, or social reality. Without belonging to the order of language, they help me make sense of the world and my place in it.

For my purposes, two aspects of Lohmar's theory need to be made explicit. First, Lohmar says in no uncertain terms that his is a theory of *primate* representation that applies to all of the more than three hundred species of primates in existence, from humans to lemurs.[14] Second, he also says that primates can activate any of these representational modes in the absence of the relevant external stimuli. For example, a rhesus monkey could activate the visual mode to visualize objects that are absent from the immediate environment and thus not given through the senses. By the same token, a gorilla could activate the visual, emotive, and gestural modes to imagine a troop of gorillas grooming each other or a couple of alpha males fighting. Of course, how different primates experience these representations will depend on their state of consciousness. If they are awake, the representations will take the form of a visual imagining or a daydream; if they are asleep, the representations will take the form of a dream or nightmare.

This seems to be what the rhesus monkeys in Vaughan's experiment did. They activated the visual and emotive modes in their sleep and, consequently, experienced a sequence of visually and emotionally rich mental images. We know these images had a visual component because they prompted a specific response that the monkeys had learned to generate following visual stimulation (namely, bar pressing), and we know they had an emotional component as well because they elicited physiological and physiognomic indices of arousal, such as deep breathing, flared nostrils, and facial grimaces. Surely, one could describe this behavior in accordance with Morgan's canon by saying that their facial muscles "twitched," but this simplest possible explanation would not begin to capture the behavior's complexity. A more reasonable description would be that the monkeys were re-enacting a waking experience in their dreams. Maybe they dreamed that they were being subjected to the same type of isolation as in their waking state, and maybe this re-enactment brought with it the same negative emotions connected to the original experience: the stress of anticipating a shock to the feet, the claustrophobia of the sensory deprivation chamber, the frustration of being held captive in a laboratory, and so on.[15] Either way, this example highlights the affinities between two denizens of Thomas's spectrum of mental imagination: hallucinations (which Vaughan was looking for but did not find) and dreams (which he found but was not looking for).

Dreams, Pretense, and Fantasy

In *On the Evolution of Conscious Sensation, Conscious Imagination, and Consciousness of Self*, the psychologist Robert Kunzendorf discusses Vaughan's research and, like Lohmar, concludes that the dreams of other primates are probably on the same continuum of activity as other mental acts that also redirect primate intentionality from the solid realm of the real to the more amorphous one of the possible, such as waking imaginings. Pointing to the periodicity and the REMs displayed by the rhesus monkeys in Vaughan's study, Kunzendorf reminds us how hard it is to conceptually pry dreams and waking imaginings apart:

> It is important to note that both periodicity and eye movement are statistically associated not only with dreaming but also with wakeful imaging. Periodicity research by Wallace and Kokoszka (1995) indicates that the vividness of wakeful visual imagery waxes and wanes throughout the day, possibly in synch with the dream cycle and with the nervous system's ultradian rhythm. Eye-movement research by Laeng and Teodorescu (2002) indicates that the human eye's movements during wakeful imaging of a previously seen stimulus tend to "reenact" its movements during initial perception of the visual stimulus. Furthermore, research by Sima, Schultheis, and Barkowsky (2013) reveals that people exhibit spontaneous eye movements as they solve spatial

> problems, but only if they construct visual images while solving problems, not if their spatial thinking remains imageless. Likewise, we expect that warm-blooded animals not only tend to dream during REM sleep but also can image during wakefulness.[16]

Kunzendorf's point is that mental acts that initially strike us as having almost nothing in common with one another may, upon further investigation, turn out to share crucial cognitive and physiological similarities. For instance, there may be a lot of differences among dreaming, waking imagination, and problem-solving, yet all these activities are exercises in "visual construction" that depend on the mind's spontaneous generation of images, a process that tends to be periodic and often leads to REMs.

Kunzendorf argues that animals partake in visual construction in all sorts of ways, and he uses playful pretense as an example. Playful pretense occurs when animals behave "as if" something were the case when it is not.[17] He cites two cases from primatology: a report of apes using inanimate objects as pretend toys, and another of an ape named Panbanisha who liked to pretend-eat the blueberries in a photograph. Here is how the primatologists who observed this second case describe Panbanisha's behavior:

> Panbanisha "eats" directly off of a picture of blueberries. She places her mouth onto the photograph, closes her lips while touching them to the picture, raises her hand

> and makes mouth movements as if chewing. After a few repetitions of this behavior, Panbanisha then picks "blueberries" off of the picture with her finger and "eats" them off her fingers—extending her mental representation of the present blueberries into visible space (i.e., away from the picture and her mouth).[18]

To grasp the complexity of this performance, consider what Panbanisha had to do in order for her subterfuge to succeed as pretense. First, she had to treat the blotches of ink on the photograph as real blueberries, thereby establishing a set of mapping rules between the real and the imaginary. Then, she had to mentally project the physical properties and causal powers of real blueberries onto the irreal blueberries in the photograph. She then had to suspend her normal way of interacting with the photograph, thereby transfiguring her relationship to reality. She had to stop relating to the photograph as a photograph and treat it as a placeholder for the irreal in the real, which is to say, as a secret passage through which the imaginary could sneak into, and momentarily become part of, her world. Finally, she had to shuttle back and forth between the real and the imaginary by performing the right behaviors in the right way and in the right order (grabbing the imagined berries, bringing them to her mouth, eating them, and then grabbing some more). In other words, she had to work out an imaginary scenario in her mind's eye and convincingly translate it into reality. Just as the dreams of the rhesus monkeys in Vaughan's study involved visual,

FIGURE 10. Primates engage in myriad forms of make-believe. Here, Panbanisha pretends to eat blueberries from a magazine photograph. Her performance involves muscular, gustatory, and visual imagery.

tactile, and emotional imagery, Panbanisha's phantasmagoric performance could succeed only by incorporating a host of "tactile-muscular, gustatory, and visual imagery."[19]

Primatology is rife with examples like this one. In an article titled "Log Doll: Pretence in Wild Chimpanzees," the Japanese primatologist Tetsuro Matsuzawa recounts the story of Jokro, a two-and-a-half-year-old female chimpanzee living in the wild in Bossou, Guinea, who one day became terribly sick with a respiratory disease. Her mother and sister took turns caring for her, and Matsuzawa noticed that when it was the mother's turn to take care of her ailing daughter, Jokro's sister (a healthy, adult female named Ja) would carry around a "log doll," caring for it as if it were her little sister. "It appeared that Ja was pretending to take care of her sick sister, using a log doll, just as she had witnessed her mother do."[20]

After a few weeks, Jokro's condition deteriorated so drastically that she could no longer hold herself up or even hold on to her mother. Still, her mother carried her everywhere until Jokro was little more than a bundle of arms and legs hanging lifelessly from her mother's back. In his account of the event, Matsuzawa focuses on Ja's use of the log doll in the days leading up to her sister's death as an example of animal pretense, but there is another candidate for pretense in the story: the behavior of Jokro's mother *after* her daughter's passing. Matsuzawa reports that Jokro's mother continued carrying Jokro's dead body on her back for two weeks after her daughter's death. It seems that she was not ready to face the death of her child and, in the throes of grief, continued to act "as if" she were still alive.[21]

I admit that using chimpanzees to theorize about pretense may be low-hanging fruit. As psychologist Robert Mitchell has observed, great apes are the only animals that scientists and philosophers will credit with capacities that involve imagination, such as pretense:

> Scientists are often hesitant to describe animal activities as pretense for fear of being accused of anthropomorphism. As a backlash against the psychological over-

> interpretation of animals' activities following Darwin, psychological behaviorism directed scientists to focus on "behavior," which referred to animals' movements relatively denuded of psychological interpretation. Although many psychologists have moved away from strict behaviorism, they are still expected to be skeptical toward any implication of complex psychology in animals [...] Most scientists assume nonpretense options, or express ambivalence, likely as a result of the aforementioned fear of accusations of overinterpretation. The same occurs in philosophy. Only with great apes are scientists comfortable attributing pretense.[22]

Sadly, even in the case of great apes, holdouts remain.[23]

In an article about the history of research on pretense in children and animals, Mitchell explains that pretense behaviors were part of scientific discourse throughout the nineteenth century but fell out of vogue (much like dreams, in my own view) at the turn of the century when new theories skeptical of animal mentation swept through the natural sciences, including biology and psychology.[24] Recently, experts have returned their attention to many of these behaviors as they have come to the realization that they are too widespread and too intentional to be skeptically dismissed. From elephants who engage in dishonest signaling to fool dominant males, to birds who feign wing injuries to lure away predators, to dogs who mock fight with one another as a form of social play, to dolphins who pretend to smoke as a way of mimicking the humans around them,[25] researchers now accept that nature is no stranger to the art of make-believe.

What intrigues me about these cases of intentional trickery and deception on the part of animals is that, again, they elucidate the connection between playful pretense and other acts of imagination, such as dreams, hallucinations, and mind wanderings.[26] Of course, we need to be cautious, since the last thing we want to do is to conflate these acts with one another. Panbanisha did not dream of blueberries any more than Jokro's mother hallucinated her daughter's death. Still, there seems to be an incontestable through line connecting these otherwise seemingly unconnected phenomena, which is what I am calling "the zoological imagination." This imagination, which is pervasive in the animal kingdom, is what unites the dreams of the octopuses, dogs, and cats we encountered in chapter 1 with the hallucinations of the rhesus monkeys and the phantasmagoric performances of the apes we just discussed. This imagination is also what connects these dreams, hallucinations, and performances to the daydreams (or mind wanderings) that neuroscientists have recently discovered in rats.

DREAMING BY NIGHT, DREAMING BY DAY (CASE STUDY 2)

In my discussion of Lohmar's theory of primate representation, I mentioned that many scientists and philosophers believe that thought is necessarily linguistic. Fortunately, this idea is slowly receding from view and making space for new theoretical frameworks that no longer assume that thought rests on the embellished laurels of language. In *Thought Without Language*, the neuropsychologist Lawrence Weiskrantz reflects upon the limits of this idea and warns that as long as we believe that language has thought on a leash, we will continue to be stunned by the mental complexity of living beings who do not meet whatever arbitrary bar we set for "possessing language." This includes people with brain damage who have lost their linguistic abilities but who nonetheless retain various levels of cognitive performance, infants who have not yet mastered a language but who nonetheless form thoughts about the world, and animals who will never have propositional language but who nonetheless can surprise us—and each other—with their sensitivity, curiosity, and acumen.

The ethologist Marc Bekoff and the philosopher Dale Jamieson say this about animals and the possibility of nonlinguistic thought:

> It may be possible to suppose that thought requires representation, but representation may not involve language [. . .] The failure to represent linguistically may affect or even limit the beliefs and desires of language-less creatures, but it is hard to see why it should prevent them from having beliefs and desires altogether, or from using cognitive maps.[27]

Like them, I see no reason why animals should need the grace of language to form mental representations, whether they be beliefs, desires, or cognitive maps. However, if we want to get to a place where thinking finally appears as something other than language's pale and inchoate shadow, we will need to explain how thought might arise without language. What capacity—or capacities—might step in to play the role traditionally assigned to language in the thought-formation process?

Imagination comes to mind.

Rodent Cognitive Maps, Revisited

Take cognitive mapping. Cognitive maps are internal mental representations of space that enable animals to travel between locations in a calculated and intelligent manner. More than a matter of linguistic mastery, I argue, cognitive mapping may be a matter of imaginative skill. Animals create, maintain, and update their cognitive maps (sometimes referred to as "Tolmanian cognitive maps," after Edward Tolman, the psychologist who discovered them) by flexing their imaginative muscle and activating something akin to Lohmar's scenic-phantasmatic mode of representation, which allows them to mentally represent scenes that do not match their immediate surroundings.

To understand how the imagination might be implicated in cognitive mapping, we need to bear in mind the philosophical distinction between *reproductive* and *productive* imagination. Reproductive imagination refers to the act of remembering images of people, places, or objects from memory, while productive (sometimes also known as "constructive") imagination refers to the act of using previously experienced images of people, places, or objects to create entirely new images that go beyond what was originally experienced. In my view, animals arrive at a unified and coherent mental representation of space by doing two things: (1) storing previous spatial experiences as images in short- and long-term memory, and later recalling these images in the absence of the relevant external stimuli (*reproducing*), and (2) synthesizing new spatial possibilities that exceed what was originally given to them directly through the senses (*producing*). This combination of storage and synthesis culminates in the generation of cognitive maps that guide animals through the space they inhabit—maps that, in my interpretation, are imaginative through and through. Given that Tolman introduced the concept of cognitive mapping while working with rats, let us return for a moment to these friendly, social creatures and consider recent findings in hippocampal research attesting to the central role of the imagination in the construction of Tolmanian cognitive maps.

In chapter 1 we saw that rats occasionally replay past spatial experiences in their sleep using specialized cells in the hippocampus. But EEG recordings have shown that rats also engage in replay *while awake*, especially when they take breaks from running a maze. When rats are exploring a maze, all their attentional resources are allocated to finding their way through it. But as soon as they are given a chance to relax, their minds start straying from their immediate surroundings. As their minds wander, they mentally replay all kinds of spatial sequences, including sequences they have personally experienced (reproductive imagination) as well as sequences they have never experienced (productive imagination). To understand what is going on in these moments of "waking replay," we need to look at several key developments in neuroscientific research on hippocampal function.

The story begins in 2006, when David Foster and Matthew Wilson from the Picower Institute for Learning and Memory at MIT discovered that, while awake, rats mentally replay spatial experiences that had recently taken place in their current physical location. What stood out for Foster and Wilson was that, unlike the experiences rats replay in their sleep, the ones they replayed while awake were "reversible." During sleep, spatial sequences are always replayed in the same order in which they were originally experienced. Trajectory X-Y-Z is replayed as X-Y-Z. But in the waking state, especially when the rats came to a "pause in exploration," trajectory X-Y-Z could be replayed either as X-Y-Z or Z-Y-X.[28] In other words, it could be replayed forward (from their current position to another destination) or backward (to their current position from another point of departure). They called this "reverse replay."[29]

This may not sound notable, but the rats in Foster and Wilson's laboratory never traversed the space in question backward. They only traversed it forward. Yet, they occasionally replayed it backward. This means that waking replay is more than a passive recapitulation of the past. It is a creative process that enables rats to envision new experiences, experiences that *could have been*. In their review of this research, Thomas Davidson, Fabian Kloosterman, and Matthew Wilson explain, "These results suggest that replayed trajectories represent the set of *possible* future or past paths linked to their current position rather than actual paths [they have taken]."[30] During reverse replay, these creatures performed a tremendous cognitive feat: they made a modal jump from the given to the imagined, from the actual to the possible. They envisioned something that never took place.

In 2009, Mattias Karlsson and Loren Frank, two neuroscientists from the Center for Integrative Neuroscience at University of California, San Francisco, took our understanding of waking replay to a new level by showing that rats also engage in what they called "remote replay," a peculiar form of replay during which animals reenact spatial sequences not linked to their current position in a maze. Karlsson and Frank found that when rats take a break in one section of a maze, they can replay a spatial sequence from a different section of the maze, even if there is nothing in their present location that could remind them of it. In remote replay, rats cognitively dissociate from their immediate environment and "replay experiences of one place while awake *in a different space*."[31] Here, the link between replay and the local environment is severed as rats re-create past experiences in the absence of direct perceptual access or mnemonic aids.[32] Given that neuroscientists often define imagination as attention to things that are absent from the local environment, remote replay can reasonably be construed as an inherently imaginative act.

We find an even more compelling connection between replay and imagination in the work of the neurologist Anoopum Gupta. In 2010, Gupta teamed up with two experts from the University of Minnesota to study rat replay. After presenting a group of rats with a maze containing multiple pathways, they observed that when the rats paused at a junction of the maze, they engaged in both reverse and remote replay, but they also did something else: they replayed spatial sequences they had never experienced before—not in the flesh, at least. Gupta and his colleagues report the case of a male rat, referred to only as "R135," who repeatedly replayed a spatial sequence that he had never experienced and that *did not exist*. Analysis of the structure of replay revealed by electrophysiological recordings showed this sequence to be an imagined amalgam of two real paths, a route leading from the rat's position to another point by means of an imaginary shortcut. The construction of this "never-experienced novel-path sequence," the authors say, is evidence that "the relationship between replay content and past experience may not be straightforward."[33] Replay is not limited to what was; it presents what *could be*. Gupta and his collaborators go so far as to say that this imaginary replay shows that rats may be capable of "self-projection," which is "the ability to consciously explore the world from different perspectives."[34] This kind of "perspective-taking," which is associated with empathy and theory of mind, is one of the cognitive abilities that scientists and philosophers have historically used to champion the uniqueness and superiority of *Homo Sapiens*.

What all this research on hippocampal function teaches us is that mainstream accounts of mental replay need to be retooled. For most of the twentieth century, psychologists and neuroscientists assumed that the only function of replay was to consolidate short- into long-term memory. The idea was that animals replay events from their past as a way of strengthening their cognitive mastery of them. But this account is incomplete. Replay also contributes to the "active construction of a Tolmanian cognitive map."[35] By replaying scenes from their past, including scenes connected and unconnected to their present location as well as scenes imagined rather than experienced, rats create global mental representations of their environment. They create cognitive maps that guide their behavior. These maps are not built solely through the accumulation, consolidation, and sedimentation of experience. They are built through a fusion of memory and imagination, through the replay of experienced actualities and imagined possibilities. They are built, that is to say, through the delicate dance between reproductive and productive imagination. As Davidson, Kloosterman, and Wilson note, every cognitive map is, at its core, an "imaginary map."[36]

Thinks like a Rat, Daydreams like a Rat

What are rats doing during moments of waking replay? There are two possible interpretations, both equally fascinating.

One is that they are thinking. James Knierim, a neuroscientist from Johns Hopkins University, argues that since some instances of waking replay are "divorced from any explicit sensory input," they qualify as acts of thought. "These reactivation events are a correlate of the rat's 'thinking' [. . .] about other parts of the track and about its recent experiences at locations other than its current position."[37] He then adds: "Could the nonlocal representations of past experience shown in the rodent studies be a precursor of the hippocampus-dependent ability of humans to imagine completely novel experiences?"[38] I do not see why not, even if I may balk at Knierim's characterization of rat imagination as a precursor to human imagination rather than as a fully developed evolutionary reality of its own.

A recurrent finding gives added weight to this interpretation. Several experiments have revealed that the longer rats rest, the more complex waking replay becomes. When they only rest for a few seconds, rats mostly replay sequences con-

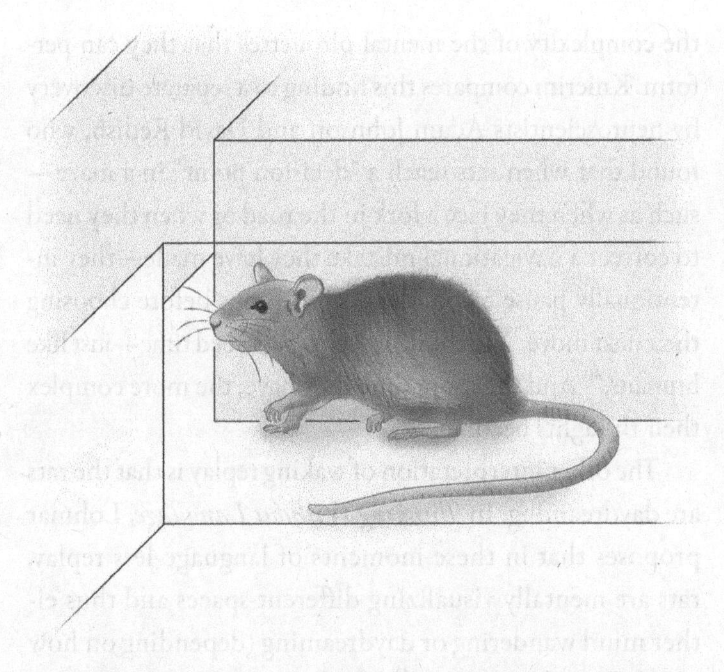

FIGURE 11. When rats come to a forked path, they pause to think. During these pauses, they replay scenes from their past as a way of assessing and evaluating the options available to them in the present. The neuroscientist James Knierim interprets these pauses as moments of thinking, while the philosopher Dieter Lohmar views them as instances of daydreaming.

nected to their current position in space, trajectories to and from wherever they happen to be at that moment. But when they rest for longer periods of time, they begin reenacting sequences that demand more complex cognitive maneuvers, such as sequences unconnected to their current physical location or sequences they have never experienced. There seems to be, then, a clear relationship between the amount of time rats have at their disposal during these periods of relaxation and the complexity of the mental pirouettes that they can perform. Knierim compares this finding to a separate discovery by neuroscientists Adam Johnson and David Redish, who found that when rats reach a "decision point" in a maze—such as when they face a fork in the road or when they need to correct a navigational mistake they have made—they intentionally pause to ponder their options before choosing their next move.[39] To think, it seems, rats need time—just like humans.[40] And the more time they have, the more complex their thoughts become.

The other interpretation of waking replay is that the rats are daydreaming. In *Thinking Without Language,* Lohmar proposes that in these moments of language-less replay, rats are mentally visualizing different spaces and thus either mind wandering or daydreaming (depending on how one defines these terms).[41] For quite some time now, neuroscientists and philosophers have been busy unpacking the behavioral, neural, and phenomenological overlaps between dreams, daydreams, and mind wanderings, and the growing consensus is that these states are more similar to one another than previously supposed.[42] It would not be surprising if a neurocognitive mechanism such as hippocampal replay were implicated in all of them. If so, some of the differences between these phenomena could be partly explained by the metabolic state of the organism at the time this mechanism is activated. When hippocampal replay occurs during sleep, rats dream. When it occurs during the waking state, they either mind wander or daydream.

Understood in this way, hippocampal replay is yet another illustration of how imagination brings disparate mental states under the same roof.

THE MUSIC OF THE SOUL

When we consider rhesus monkeys frantically pressing a bar in their sleep, apes pretending to eat fruit from a photograph, or rats visualizing irreal paths through space, we are led to conclude that the imagination may be unique and it may be human, but it is not uniquely human. It is a zoological, rather than anthropological, reality. It is the music of the animal soul.[43]

If we listen carefully to its rhythms, to its peaks and valleys, this music can expand and transform our habitual modes of perceiving animals and relating to them. In the words of the psychologist Thomas Hills:

> If [animals] can imagine, one can't help but wonder what kind of cognitive system is required for such a feat. What other implications do imaginings create? Does an animal that can deliberate about its future enjoy some modicum of free will? Might an animal that imagines also know that it is imagining? Might it know the difference between the real and the imagined? Might it, therefore, know the difference between its real self and what it could be?[44]

I do not know how we should answer these questions, but acknowledging creativity and imagination as animal characteristics changes the rules of the game. Just as dreams open onto the horizon of the imaginary, Hills's remarks suggest that the imaginary itself opens onto an even wider horizon in which animals are seen as containing hitherto unimagined depths—cognitive, emotional, and, as we will see in the next chapter, even moral depths.

CHAPTER 4

The Value of Animal Consciousness

Consciousness is that through which we know what's around us, what's on our minds, and what we mean; it is that by which we have minds at all; and it lies at the heart of our concern for ourselves and for others. And if that doesn't make it important, what would?

—CHARLES SIEWERT[1]

[Consciousness] makes life worth living.

—DAVID CHALMERS[2]

MINDING ETHICS

Beginning in the 1970s, the ethologist Donald Griffin caused an uproar in the life sciences by claiming that animals are minded beings who are aware of their surroundings. In books such as *The Question of Animal Awareness* and *Animal Thinking*, Griffin developed a theory of animal cognition that held that animals internally represent the external world and use these representations to navigate their environment in flexible ways. If consciousness exists in humans, he reasoned, it must exist in other species too.[3] With this seemingly simple conditional statement, Griffin fathered cognitive ethology and, less directly, the subfield of academic philosophy known as "the philosophy of animal minds" or "the philosophy of animal consciousness." In this book, I have suggested that we can make progress in this subfield by zooming in on the side of animal consciousness that surfaces during sleep, when the animal mind, as Gay Luce says, "speaks only to itself."[4]

Yet, even if one accepts my claim that dreaming is a portal to animal consciousness, one may still wonder what difference it makes in the grand scheme of things whether animals are conscious or not. The answer is that it makes a huge difference from an ethical point of view. As the philosopher Mark Rowlands explains:

> The quickest way to deny animals moral status [. . .] is to deny them mental status; it is to deny that they are subjects of mental states or to deny that they have a mental life. Such a denial might strike the person on the street—at least if their street is in any way populated with animal life—as absurd. This, however, has not stopped many prominent philosophers from issuing this denial.[5]

Following Rowlands, I believe that attributing conscious states to animals is an inherently ethical gesture as there is a link between being conscious and having moral status—and, conversely, between being seen as lacking consciousness and being exposed to unimaginable forms of cruelty.

Still, getting to moral status from consciousness is not easy. As Marc Bekoff and Dale Jamieson have rightly noted, "One cannot immediately move from views about animal minds to views about the moral status of animals. Some important inferential connection must be established between them, and this requires argument."[6] In this chapter I build one such inferential connection with the help of dreams. Dreams have a hitherto unrecognized moral force insofar as they are expressions of what the philosopher Ned Block calls "phenomenal consciousness," which I believe grounds the moral status of living beings. Animals who dream should be recognized—*because* they dream—as members of the moral community, as fellow creatures who deserve to be treated with care, dignity, and respect.

CONSCIOUSNESS AND MORALITY

Most Western schools of moral theory begin from the assumption that we only have ethical duties toward other conscious beings. Even as they disagree about what consciousness is and which organisms have it, very few of these schools endow nonconscious entities, such as planets, rocks, or paintings, with inherent moral worth.[7] Most of them wholeheartedly embrace philosopher David Chalmers's view that consciousness "has special value," namely, moral value.

This special value stems from the fact that consciousness confers moral status to those who possess it. When we anoint a creature with the distinction of being conscious, they are transformed in our eyes from creatures who merely exist into creatures who exist *and* matter from a moral point of view. As the ethicist Mary Anne Warren explains in her book *Moral Status: Obligations to Persons and Other Living Things*:

> To have moral status is to be morally considerable, or to have moral standing. It is to be an entity toward which moral agents have, or can have, moral obligations. If an entity has moral status, then we may not treat it in just any way we please.[8]

Consciousness entitles organisms to what Warren calls moral consideration by making them matter for their own sake. To be conscious is to be some*one* rather than some*thing*, a "thou" rather than an "it."[9]

We saw in our analysis of consciousness in chapter 2 that consciousness is not a simple, homogenous unity but rather a complex phenomenon with various types. This raises a crucial question: Which of these types is responsible for the conferral of moral status, or do all types of conscious awareness perform this function equally? Do we just need to be conscious to matter morally, or do we need to be conscious in a specific way?

Over the past twenty-five years, a small but thought-provoking body of philosophical work has developed this question. From its inception, most of this work has been framed through Ned Block's influential theory of consciousness, which divides consciousness into two types: "access" and "phenomenal." As a result, two major camps have emerged. Access-first theorists believe that access consciousness is the foundation of moral worth, while phenomenal-first theorists see moral status as arising from phenomenal consciousness instead. Both camps share the conviction that consciousness grounds moral value, but they disagree vehemently about which kind of consciousness does this important work. As we shall see, this disagreement boils down to competing visions of moral life: one that centers cognition, rationality, and language, and one that, taking a less cerebral approach, privileges our subjective, affective, and embodied rootedness in the world.

Block's Theory: Access versus Phenomenal Consciousness

In a paper that has become a classic in the philosophy of consciousness titled "On a Confusion about a Function of Consciousness," Block laments that debates about consciousness have been derailed by a lack of conceptual clarity. Chances are high that when experts disagree with one another about what consciousness is, where it comes from, and how it works, they may be talking about different things without realizing it.

To put an end to this confusion, Block divides consciousness into two kinds: "access" and "phenomenal." Access consciousness refers to representational mental states whose contents are available to the wider cognitive system for use in executive functions such as reasoning, decision-making, and linguistic reporting. Block defines this form of conscious awareness as follows:

> A [mental] state is access conscious [. . .] if, in virtue of one's having the state, a representation of its content is (1) inferentially promiscuous, that is, poised for use as a premise in reasoning, (2) poised for rational control of action, and (3) poised for rational control of speech.[10]

In the highbrow slang and tortured syntax of academic philosophy, this just means that a mental state is "access conscious" when we can think rationally about its contents, use them to make decisions about our behavior, and share them with others through the medium of language. For example, if I believe that there is a door at the end of the hallway, this thought is access conscious as long as I can use it to conclude that the cat may get out if I leave the door open (make an inference), walk to the door and close it (guide my behavior), or tell my partner to close the door because it is cold outside (make a linguistic report). Lots of our mental states are access conscious in this way.

Phenomenally conscious states are harder to pin down, but they differ from access conscious states in two important respects. First, they are nonfunctional. They are not tied in any essential way to the performance of any specific cognitive operations. They do not lead to inferences, voluntary action, or communicative acts. Second, as their name indicates, their content is phenomenal rather than representational, meaning that they have a determinate feeling associated with them, but they do not represent anything in the external world—not objects, not people, not places, not even states of affairs. We find ourselves "in" these states, but they are not "about" anything.

From the moment he introduced the distinction between access and phenomenal consciousness, Block knew that defining phenomenal consciousness would be an uphill battle. Because functionality and representationality belong to access consciousness, he could not define phenomenally conscious states by describing what they do or what they refer to. He got around this problem by listing examples of phenomenally conscious states, hoping this would suffice to give readers an intuitive grasp of the concept.

He drew most of his examples from the realm of the senses. "We have [phenomenally] conscious states when we see, hear, smell, taste, and have pains."[11] Why? Because there is a qualitative feeling attached to seeing a color, hearing a melody, smelling an odor, tasting a dish, or experiencing a particular kind of pain. Each of these mental states has a lived quality that we cannot convey to others, especially if they have never experienced those things on their own. How do we explain what it is like to see green to a congenitally blind person? How do we describe the bitterness that coats our mouth when we eat an unripe banana to someone who, for whatever reason, has lost all taste receptors as well as all memory of taste? How do we convey the experience of seeing in three dimensions to someone who sees only in two? As the philosopher Neil Levy explains:

> Unfortunately, it seems to be impossible to define phenomenal consciousness. The best we can do is to point to instances of it. Phenomenal consciousness is the kind of conscious such that there is something it is *like* to be phenomenally conscious. There is something it is like to taste a glass of pinot noir, to hear the opening notes of *Tristan und Isolde*, to feel the warmth of the sun on your face or to feel pain in your left knee. Each of these experiences has a distinctive phenomenal quality to it. That quality seems inexpressible. We often use metaphorical language to communicate that quality ("it is like a dull throbbing ache"; "a sharp stabbing pain"; "a rich lush red"), but when we do so we seem to rely on our shared experience with such phenomenal qualities to calibrate our talk.[12]

If you have never tasted pinot noir, I can only describe it to you either metaphorically (for example, by telling you, "It's a strong, almost unctuous, pulp-free berry juice with a tannic profile that coats the mouth") or indexically (that is, by making you taste one pinot after another until you get the general idea and, probably, a serious buzz). But even with all the poetic license in the world, my metaphors will necessar-

FIGURE 12. The philosopher Ned Block's theory of consciousness differentiates between access and phenomenal consciousness. Phenomenally conscious mental states, such as the taste of red wine, have a qualitative feeling essentially attached to them. This feeling does not serve any cognitive function or represent any internal or external state of affairs. Other examples of phenomenally conscious states are seeing colors, hearing melodies, and feeling pain.

ily fall short because there will always be a gap between my first-person experience of red wine and my descriptions of it. That gap is none other than its *taste*, which can only be experienced in the flesh.

Pain is another example. If I tell my partner to come to the living room and he stubs his toe on a piece of furniture on the way, I can ask him what happened, where it hurts, and even how badly it hurts, but it would be strange for me to ask him, "What is your pain *about*? What does it *represent*?" By its very nature, pain is a nonrepresentational mental state that does not refer to anything in the world. It is not "about" anything. Rather, my partner is "in" pain. Suddenly, he senses an intensity at the end of his foot that distracts him from everything else going on around him. His entire being is overwhelmed by the feeling—the phenomenology—of hurting.

In *The Mystery of Consciousness,* the philosopher of mind John Searle observes that even if a lot of things happen concurrently the instant my partner stubs his toe (his neurons fire, he lets out a cry, he attends to the injury, and so on), the most important is that he *feels* an unpleasant sensation. This subjective feeling is what makes pain "phenomenal" rather than "representational," because even if there are cognitive aspects to my partner's experience of pain, there is something about it that is irreducible to cognitive processing. Yes, my partner can explain to me what happened, where it hurts, and how badly, but he cannot transfer his pain to me. This pain is his. As much as I may want to, I cannot bear it for him. His pain is inalienable, incalculable, and ineffable. As Searle puts it, "The essential thing about the pain is that it is a specific internal qualitative feeling. The problem of consciousness in both philosophy and the natural sciences is to explain these subjective feelings."[13]

In sum, access consciousness involves mental representations and cognitive functioning, while phenomenal consciousness involves precognitive feelings and lived experience. The former involves the movement of information through our cognitive apparatus, enabling rational thought, behavioral control, and linguistic expression. The latter is the raw feeling of being in a state that is functionally and representationally inert yet phenomenally and experientially rich. When I am in pain, I am not representing some aspect of the world or performing an elaborate cognitive function. I am living through an experience, an experience of bodily intensity. The same is true when I drink a pinot, when I hear leaves rustling in the yard, and when I smell an unpleasant odor wafting up from the sewer grates in my neighborhood.

Lest we lose ourselves down the rabbit hole of academic squabbles concerning the nuts and bolts of Block's theory (which can get quite technical), let's return to our main point, which is that this theory has framed the broader academic debate about the moral value of consciousness. Block's assertion that there are two kinds of consciousness has led many to wonder which of these is required for moral status. Do we matter morally because we have reached the right level of cognitive functioning or because we have the right kind of phenomenology? What makes us matter: rationality or lived experience, access or phenomenal consciousness?

Phenomenal Consciousness, the Seat of Moral Status

Let me put my cards on the table: I believe phenomenal consciousness is what matters for moral status. In my view, what brings organisms under the aegis of morality is not their ability to think rationally, act voluntarily, or produce linguistic reports (the defining features of access consciousness in Block's account), but the fact that they have a phenomenally charged experience of the world—that they sense, feel, and perceive.

Charles Siewert, an expert in the relationship between philosophy of mind and ethics, shares this view. He uses a philosophical thought experiment to support his position. Imagine, he says, that you are given the choice to become a zombie. As a zombie, you would be functionally equivalent to your present self (that is, you would do everything you already do and nobody would know the difference), but you would also be phenomenologically depleted (you would have no conscious awareness of your surroundings, no sense of what it is like to do X, Y, or Z). In other words, zombie-you would go through the motions flawlessly, fooling even your closest friends into thinking that it is the real you, but it would have no inner life whatsoever. It could gulp down a bottle of wine and act drunk, but it would not taste the pinot. It could stub its toe on the furniture and let out a yell, but it would not feel an intensity at the end of its foot.[14]

Siewert is convinced that even if the deal were sweetened (with a huge sum of money, let's say), none of us would take the plunge because, ultimately, we do not just value the cognitive functions that we perform by being conscious, we value being conscious itself. We treasure the feeling of being phenomenally aware of our surroundings, of being alive, of taking the world in through the senses. Phenomenality is not something we are likely to surrender because it is too fundamental to who—and what—we are. Even if we were told, "Look, all things considered, zombification would be a net benefit because, as a zombie, you would never feel pain or sadness again," most of us would respond, "That is exactly the problem! As a zombie, I wouldn't feel pain because I wouldn't feel anything at all, and there is no value in that. I'd rather feel than not feel, even if pain and suffering figure prominently in the mix." All of us, Siewert suggests, feel in the marrow of our bones that what makes us suitable recipients of empathy is our subjective, embodied, and affective—in one word, phenomenal—anchoring in the world. This anchoring gives us moral import. It grants us moral status.[15]

There are two levels to Siewert's position. The first is that phenomenal consciousness itself has intrinsic value. Being phenomenally conscious is inherently good, even if it opens us up to the possibility of pain and suffering. The second is that that we have moral value because of phenomenal consciousness. What horrifies us about a hypothetical zombification procedure is that, as zombies, we would no longer have moral value. We would be mere zombies. As such, we would be phenomenologically and, *ipso facto,* morally bereft. More than fearing the loss of something we value, we fear the loss of our own value, the loss of our identities as beings who deserve to be treated ethically.

To these two levels we can add a third inspired by the work of the philosopher Joshua Shepherd.[16] Phenomenal consciousness is something we value and something that bestows value upon us. But it is also what makes the very act of valuing possible in the first place. It enables living organisms to inject value into an otherwise value-neutral universe.[17] A creature without phenomenal consciousness would have no lived experience of the world, no feeling of the here and now, no sense of what is positive or negative (or, consequently, better or worse). Even if such a creature could perform a wide range of cognitive functions, it could never *value*. Lacking phenomenal anchoring, it would have no basis for valuing, no ground on which to establish a preference, interest, or desire. Such a creature would have no drive to like any one thing more than another. A universe peopled exclusively by creatures of this kind would be a universe without agents who value and, thus, without objects of value—a universe with no value at all.

The salient philosophical point here is that phenomenal consciousness takes precedence over access consciousness in the realm of ethics because it, and it alone, confers moral status. Access consciousness may add cognitive and behavioral complexity to our lives, but it is not the source of our moral standing; only phenomenal consciousness is. It may modulate moral value, but it cannot generate it; only phenomenal consciousness can.

THE ACCESS-FIRST APPROACH: THE OTHER SIDE OF THE ARGUMENT

It is no secret that Western philosophers, especially moral theorists, have historically put the functions of access consciousness on a pedestal. For them, humans are entitled to moral protection because we are creatures who think rationally, behave rationally, and communicate linguistically, which is to say, because we are creatures who possess access consciousness. Without the latter, they contend, we would have no moral value; we would remain an "it" instead of a "thou."

Since this access-first approach to moral status comes in two versions, one consequentialist and the other deontological, I will discuss each in turn.

The Consequentialist Version

As a rule, consequentialists are a class of moral theorists who see the maximization of happiness and the minimization of suffering as the *summum bonum* of moral life and believe that our actions ought to be evaluated according to the extent to which they amplify or diminish the overall amount of happiness in the world. But lots of things bring about happiness, so how do consequentialists decide which ones to prioritize? On this point, there is a wide range of positions. Hedonic consequentialists argue that the only things that matter are pleasure and pain, while preference consequentialists stress that not all pleasures are equivalent and that ranking them is one of the functions of moral philosophy.[18]

Historically, consequentialists like John Stuart Mill have ranked pleasures using a cognitivist formula according to which pleasures that require complex cognitive processing (such as appreciating art, cultivating friendships, acquiring new knowledge, and developing our talents) must be ranked "higher" than those wedded primarily to the senses (such as smelling the roses, waking up from a good nap, and sunbathing at the beach). We see this formula at work in Mill's *Utilitarianism*, where he insists that even if playing pushpin (a popular nineteenth century children's game) brings someone as much pleasure as reading poetry, we should still say that poetry is an objectively "higher" good than pushpin because it is more intellectual. This is in contrast with Jeremy Bentham's claim that if pushpin and poetry make someone happy in equal measure, both are equally good from a utilitarian perspective.[19]

What does this have to do with Block's theory of consciousness? Recently, a small cohort of philosophers affiliated with the Oxford Centre for Neuroethics in England (notably Guy Kahane, Julian Savulescu, and Neil Levy) have used arguments rooted in preference consequentialism to defend an access-first approach to moral status. These thinkers concede that there is value in the experiences that phenomenal consciousness makes available to us, such as seeing colors, hearing melodies, and feeling bodily pleasure. However, inspired by Mill's famous claim that "it is better to be a human being dissatisfied than a pig satisfied; better to be Socrates dissatisfied than a fool satisfied," they contend that our preferences for these phenomenal goods pale in comparison to our preferences for the goods proffered by access consciousness, such as the ability to learn new things, solve challenging problems, make rational inferences, cultivate friendships, communicate with others, and so on. Forced to choose between the perks of phenomenal and access consciousness, they say, we would—and should—choose the latter because a life without cognitive access is not a life worth living. Again, these philosophers do not deny that losing phenomenal consciousness would be bad. They simply believe that losing access consciousness would be an infinitely greater tragedy, nothing short of a moral catastrophe.[20]

I disagree with this stance. To avoid misunderstandings, allow me to clarify that cognitive access matters tremendously for those who have it. I, too, enjoy learning new things. I, too, enjoy developing my talents. I, too, enjoy spending quality time with friends. And it is safe to say that I would suffer if these things were taken away from me against my will. That being said, the question is not whether such things have value, but whether they are the *ground* of moral status. Are these things so important that anyone who lacks them must lack moral status as a result? Access-first theorists believe so. In their eyes, creatures who lack access consciousness are morally irrelevant. Morality simply does not apply to them.

Curiously, none of these thinkers shy away from what I take to be the morally troubling implications of their view. What might they say about the moral status of animals who do not have the same level of cognitive sophistication as humans but who can nonetheless feel pleasure and pain? Simple: animals have no access consciousness and thus no right to life.[21] What about patients in a persistent vegetative state who remain minimally conscious of the world around them?

Simple: they have no access consciousness and thus we have zero moral obligation toward them.[22] What about human patients who, because of brain trauma, cannot communicate but remain conscious of their natural and social environments? To avoid accusations of misrepresenting their views, I'd better let Kahane and Savulescu speak for themselves: "Terminating these patients' lives might be morally *required*, not merely permissible."[23] Yes, you read that right. Killing people with brain damage is not only morally tolerable, but also morally required. Somehow, it would be wrong *not* to kill them.

Recall Warren's claim that moral status prevents others from treating us in any way they please. It is our first and last line of protection against the most sadistic abuses, the most senseless brutalities. By reducing moral status to cognition, access-first theorists give those who meet their cognitive requirements carte blanche to treat those who do not—the cognitively disabled, people with brain damage, and all nonhuman animals—on par with inanimate objects, as things to be used, abused, and even destroyed. Because they see cognitive access as the be-all and end-all of moral life, they commit themselves to the morally abhorrent view that anyone who does not reach a certain level of cognitive performance is disposable.[24]

Access-first theorists remind me of the neurologist Oliver Sacks's famous patient "Dr. P.," who could only see the abstract and the schematic but not the concrete and the living. They get so caught up in the cognitive scaffolding of human existence that they forget about the precognitive and prelinguistic foundation on which this scaffolding rests. They forget—or, on a less charitable reading, repress—what the French existentialist Maurice Merleau-Ponty calls "the soil" of lived experience, which is our embodied and embedded relationship to the world prior to the intrusion of reason, concepts, or language.[25]

In *Consciousness and Moral Status,* Joshua Shepherd rebukes access-first theorists for ennobling cognition to the point of degrading phenomenality. While acknowledging that all conscious states infuse our existence with value, Shepherd believes that the ones that "ensoul" us are not those that give us Blockian access, but those that put us in direct communion with reality, such as experiencing joy, being in a good mood, not feeling pain, having a moment of inner peace, feeling waves washing over our feet, enjoying the bliss of the morning sun, or giving ourselves up to the serenity of a peaceful night. He cites the following passage from William James's 1899 essay "On a Certain Blindness in Human Beings":

> Living in the open air and on the ground, the lop-sided beam of the balance slowly rises to the level line; and the over-sensibilities and insensibilities even themselves out. The good of all the artificial schemes and fevers fades and pales; and that of seeing, smelling, tasting, sleeping, and daring and doing with one's body, grows and grows. The savages and children of nature, to whom we deem

> ourselves so much superior, certainly are alive where we are often dead, along these lines; and could they write as glibly as we do, they would read us impressive lectures on our impatience for improvement and on our blindness to the fundamental static goods of life.[26]

Our moral status comes from our primordial bond with the world, where our sensations, affects, and emotions comingle with one another in an endless dance. Located within experience but outside the reach of cognition, this bond is what remains once we have peeled away the abstractions of reason and the schemas of language. It is our moral and existential bedrock.

The Deontological Version

Unlike consequentialists, who locate morality in the maximization of happiness, deontologists locate it in unconditional respect for the irrevocable dignity of others. According to them, leading a morally upstanding life requires that we respect the fundamental dignity of others by treating them as "ends in themselves" rather than as "means to an end."

Unfortunately, deontologists tend to base dignity on rationality, making moral status a function of cognition. In *Groundwork for the Metaphysics of Morals*, for instance, Kant says that our dignity depends on our "rational nature" such that only rational beings are worthy of moral esteem. Many modern-day Kantians have echoed this view and espoused an access-first interpretation of moral status, according to which our moral value resides in cognitive access (rationality) rather than phenomenality (lived experience).[27]

In several publications, the philosopher Uriah Kriegel cautions against this move. He says that we ought to think twice before following Kantians down an access-first road that leads to dubious moral conclusions. In an article entitled "Dignity and the Phenomenology of Recognition-Respect," he exposes the fault lines in the deontological approach to moral status with a thought experiment that takes this approach to its logical conclusion. He introduces us to two imaginary candidates for moral status and asks us to consider which of them deserves to be treated with moral respect. There are the "weather watchers," who are:

> conscious, feeling creatures who are incapable of any action [. . .] These are pole-like creatures who are completely immobile, rigidly stuck to the ground, but who nonetheless can sense the ambient temperature, care about it, and take great interest in it. They prefer warm weather, hope for it every morning, and are cheerful when they feel it and disappointed when they do not. They thus have a rudimentary perceptual, cognitive, and emotional life, but crucially, they have no capacity for action, and we may stipulate that their faculty of will has atrophied as a result—they experience no such states as deciding, intending, or choosing.[28]

Then, there are the self-ruling robots:

> Conversely, imagine our world contained certain end-setting automata or zombies. It is beyond doubt that much of our behavior is unconsciously driven, which seems to entail that we have many purposes and goals—including, presumably, ultimate goals, that is, ends—that are unconscious. Imagine now a creature all of whose ends are unconscious; indeed, all of its mental life, such as it is, is unconscious. It experiences no feelings or emotions, no thought processes, no bodily or perceptual sensations. Yet its unconscious life is sufficiently robust a duplicate of ours that it engages in sensible, goal-directed behavior.[29]

The difference between these two candidates is straightforward. The weather watchers have a subjective and affective experience of their surroundings but perform no cognitive functions. They possess phenomenal consciousness but not access consciousness. The self-ruling robots are just the opposite. They are rational and logical since they make decisions based on a preexisting algorithm, but they lack what the British philosopher Galen Strawson calls "mental reality."[30] They possess access consciousness but not phenomenal consciousness.

Between the watchers and the robots, then, who are the true "Kantian dignitaries"? Kriegel says that the dignitaries in this story are the weather watchers since they are sentient beings who care for the world, take an affective interest in it, and have a perspective on it. Compared to them, the self-ruling robots are machines without an inner life. The robots do not feel. They do not flourish, suffer, or yearn. They do not even live or die. Barred from all phenomenality, they are exactly what animals have historically been taken to be: mechanical contraptions without a Bergsonian élan vital, inanimate objects that one can assemble, disassemble, and reassemble at will. One does not need advanced training in moral philosophy to understand that it is the weather watchers who matter from a moral point of view.

Unfortunately, this answer is not available to defenders of the access-first approach. Those who posit access consciousness as the foundation of moral status have no choice but to argue that (i) it would be morally permissible for us to instrumentalize the weather watchers because they do not have access consciousness; and (ii) it would be morally impermissible for us to do the same to the self-ruling robots since, technically, they are rational beings who rule themselves. Yet, this position is morally confused given that when it comes to these robots, as Kriegel says, "there is nobody home."[31] By contrast, the weather watchers are sentient beings who make a compelling moral claim on us. Even stuck to the ground, they answer our moral call.

Kriegel finds the idea that "modern-day Roombas"[32] could have more moral value than creatures who yearn for a better (and warmer) tomorrow so intolerable that he wonders whether anyone, even Kant himself, could truly embrace it. If Kant had encountered the lovely weather watchers at the height of the Enlightenment, perhaps he would have fine-tuned his position and clarified that we may need a rational nature to think morally (to be agents of morality), but not to have moral status (to be objects of moral consideration).[33] Kriegel even reconstructs the logical structure of Kant's moral philosophy along the lines of this imagined possibility as follows:

1. an action is morally right only if it treats dignitaries (creatures possessed of dignity) as ends rather than means;
2. all and only phenomenally conscious creatures are dignitaries; therefore,
3. an action is morally right only if it treats phenomenally conscious creatures as ends rather than means.[34]

In this reconstruction, phenomenality, not cognition, does the heavy lifting. Phenomenality determines who is a suitable recipient of empathy. Run-of-the-mill Kantians may decry this reconstruction as misguided given the value they ascribe to rational cognition, but Kriegel points out that unless we are prepared to let lifeless robots claim moral priority over living organisms who experience feelings such as joy, hope, and disappointment, we must disabuse moral theory of the idea that only the cognitive layers of existence harbor moral value.

One final thing to emphasize. Kriegel locates the foundation of our moral status in what he sees as the most important feature of our phenomenal experience of the world: its radical inaccessibility to others, its unconditional mine-ness.

Nobody can appropriate my phenomenal experiences, just as I cannot appropriate anyone else's. Yet even though we do not have access to the contents of other people's phenomenal experience, we do have access to this lack of access. When I encounter another person, I immediately and intuitively grasp the inaccessibility of their inner lives and become conscious of their being conscious, which leads me to experience that person as morally inviolable, as someone who commands from me the most solemn moral respect.[35]

Kriegel uses a mundane example. If I walk into a café and look around, I will perceive all sorts of objects: a painting, a man sitting in the back, an espresso machine, a fire extinguisher, the menu on the board, tables, chairs, and so on. Since all these objects are denizens of my perceptual field, they revolve around me like planets around the sun. Nevertheless, there is one object among them that stands out as categorically different from the others, and that is the man sitting at the other end of the café. Surely, he is an object of my perception as much as the fire extinguisher in front of him and the painting behind him. But what sets him apart from these inanimate objects is that he has a double identity. He is an object of my perception and a perceiving subject in his own right. He is a planet that orbits me as well as a sun "complete with an army of stars and planets orbiting *him.*"[36] Phenomenologically speaking, he orbits me as much as I orbit him.

The crux of Kriegel's argument is that at no point do I ask myself, "Is the man sitting over there a something or a someone, a planet or a sun?" As soon as I step into the café and see him, I feel his heat. I feel it even before I have a chance to doubt it. In making this claim about the structure of intersubjectivity, Kriegel follows a long tradition of phenomenologists, including Max Scheler, Ludwig Wittgenstein, Edith Stein, Maurice Merleau-Ponty, and Emmanuel Levinas, who believe that we do not reason about the moral status of others; we perceive it *sui generis.*[37] We see it in their way of being-in-the-world much like we see someone's emotions in their physiognomy—the joy in the laughter, the anger in the furrowed brow. Like emotions, the inner light of others is not inferred. It is perceived. And this perception is the root of empathy. On some readings, it is empathy itself. One of my favorite definitions of empathy comes from the German philosopher Edith Stein who, in her 1917 book *On the Problem of Empathy*, defines it simply as the mode in which one consciousness perceives another. This is also Kriegel's view.

But what is it about the man in the café that I experience as morally salient? Is it his body, which I perceive as anatomically analogous to mine? Is it his face, which my brain has evolved to recognize as human and familiar? Is it his use of language, which I read as a sign of his moral value? No, no, and no. What evokes in me a feeling of respect for his dignity is the inaccessibility of his experience, my experience of his inner world as an ontological impasse *for me*. The man in the café carries in his breast an entire universe that evades me, an infinity that "resists annexation."[38] When I step into the café and spot him across the room, I immediately grasp that I am in the presence of deadbolt I cannot break, a vault I cannot open. This resistance *is* his dignity, and my experience of his dignity is what holds me ethically accountable to him.

Counterintuitive as this may sound, Kriegel believes that phenomenality grounds moral value not through a presence that fills it, but through an absence that haunts it, an absence that, in effect, constitutes it. It is not that something positive about others is presented in my experience of them. It is that something is given in my experience of them as positively un-presentable—namely, their consciousness, the almost absurd fact that they *are*.

THE MORAL FORCE OF DREAMS

For a long time, people have interpreted dreams as safeguarding, and thus possibly betraying, our most intimate secrets. Plato warned in *The Republic* against the liberation of "unnatural desires" in dreams,[39] while in the *Confessions* a troubled Augustine anguished over the sordid dreams of fornication that visited him in his sleep. Is dreaming of sin, the Bishop of Hippo fretted, in and of itself a sin in the eyes of the Almighty? Fifteen centuries later, Henry David Thoreau similarly wondered whether dreams might reveal our true moral character. In *On the Banks of the Concord and the Merrimack*, he writes:

> In dreams we see ourselves naked and acting out our real characters, even more clearly than we see others awake.

> But an unwavering and commanding virtue would compel even its most fantastic and faintest dreams to respect its ever wakeful authority; as we are accustomed to say carelessly, we should never have *dreamed* of such a thing. Our truest life is when we are in dreams awake.[40]

Augustine wiggled his way out of his dilemma by differentiating between "happenings" that befall us (such as dreaming of having sex) and "actions" that we perform with conscious intent (such as having sex). Less terrified by the concept of hell, Thoreau came to a different, squarely anti-Augustinian, conclusion. The actions we perform in our dreams reflect the habits we cultivate while awake, making them genuine extensions of our personalities. What happens in our dreams is a measure of our moral fiber, a "touchstone of our character."[41]

While I find these ways of thinking about dream morality intriguing from the standpoint of the history of ideas, I believe that Plato, Augustine, and Thoreau approached the issue from the wrong angle. Despite their differences, they all operated under the assumption that the moral force of our dreams resides in their content, and that mastering this force is a simple matter of judging this content as either "moral" or "immoral." In this respect, they were mistaken. They were right in thinking that dreams have a moral force, but wrong in presuming that this force resides in their content. For me, this force resides elsewhere, namely, in the elemental nexus between dreams and phenomenality. I already argued above that phenomenality is the seat of moral value, which means that phenomenally conscious mental states confer moral status to the organisms who experience them. Next, I want to argue that dreaming is one such phenomenal state. In fact, dreaming may very well be "the" phenomenal state par excellence. As such, it is replete with moral force.

Dreams as Phenomenal States

In his 1995 article, Block sought to establish the independence of phenomenal consciousness from access consciousness by looking for an example of a conscious state that is rich in phenomenal content but that is not cognitively accessible. He claimed to have found such a "purely" phenomenal state in the experience of research subjects who participated in partial-recall experiments conducted by the cognitive psychologist George Sperling in the late 1950s.[42] While I have no objections to Block's interpretation of Sperling's research, I believe that Block missed a more obvious and compelling example of pure phenomenality: dreaming. Dreams are mental states that present us with phenomenal content while screening off cognitive access. In them, we experience all the subjective states that Block associates with phenomenal consciousness (that is, images, sounds, smells, pain) without the level of executive functioning that defines access consciousness (rational thought, behavioral control, linguistic reportability).

The philosopher of cognitive science Miguel Ángel Sebastián is the most vocal proponent of this interpretation of dreams as phenomenal experiences purged of cognitive access. Turning to the neuroscience of dreaming, he explains that in the dream state we experience, at the subjective level, a stark reduction in "voluntary control and reflective thought" that is matched, at the neuronal level, by an equally stark drop in activity in the area of the brain that brings about cognitive access: the dorsolateral prefrontal cortex (dlPFC).[43] The dlPFC is considered vital for high-level cognitive functions such as planning, strategizing, and paying attention. It is an important component of "working memory," which allows us to temporarily store and manipulate information in the present without losing track of our objectives. The dlPFC, therefore, appears to be closely intertwined with access consciousness. During dreaming, however, it shuts off, suggesting that it is not essential for a perfectly normal, perfectly full dream life. In "Dreams: An Empirical Way to Settle the Discussion Between Cognitive and Non-Cognitive Theories of Consciousness," Sebastián writes:

> There is general agreement that we dream, though most surely not exclusively, during the REM phase. In this phase, some areas are even more active than during wakefulness, especially the limbic areas. In the cortex, areas that receive strong inputs from the amygdala, such as the anterior cingulate and the parietal lobe, are also activated; this aids in explaining the highly emotional

> component of dreams. In contrast, the rest of the parietal cortex, the precuneus and the posterior cingulate are relatively inactive. More interestingly for my present purposes, there is a selective deactivation (compared with during wakefulness) of the dlPFC [. . .] Taking into account the role of the dlPFC in cognitive access, these results suggest that we lack cognitive access during the REM phase of sleep as it does not seem plausible that another brain area plays the role of dlPFC in cognitive access during sleep. However, we dream during this phase; if subjects were wakened from this phase of sleep and asked whether they have dreamed, they reply positively at least 80% of the time. And dreams are conscious experiences, aren't they?[44]

Indeed they are. As the empirical evidence suggests, ordinary dreams are experiences of phenomenal consciousness in the absence of access consciousness.[45] The only time things look different is when we *lucid* dream. During lucid dreams, the dlPFC is reenlisted into the dream-generation process, bringing about a sudden increase in our subjective experience of voluntary control and reflective thought.[46]

Across the waking-sleeping-dreaming spectrum, then, there is a clear logic to how the dlPFC operates. It is "on" when we are access conscious and "off" when we are only phenomenally conscious. This has prompted Sebastián to speculate that the dlPFC may house the holy grail of contemporary neuroscience: the neural correlates of access consciousness.

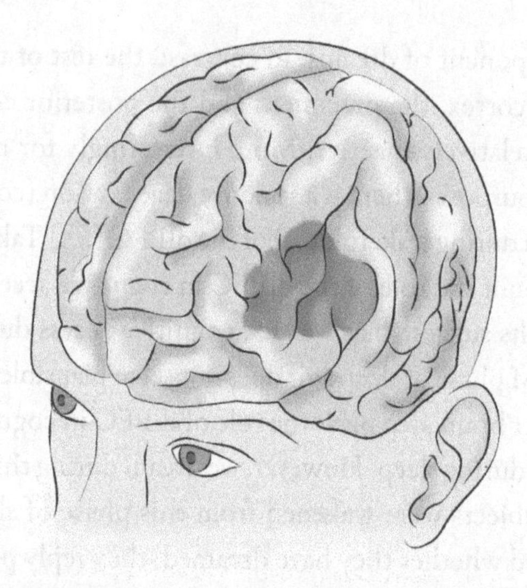

FIGURE 13. The dorsolateral prefrontal cortex (here shaded in gray) is a region of the frontal lobes thought by many neuroscientists to contain the neural correlates of access consciousness due to its involvement in rational thought and executive control. This region is deactivated during non-lucid dreams, which supports the view that these dreams are phenomenally conscious, but not access conscious, mental states.

Sebastián's theory suggests that Block was correct in thinking that access and phenomenal consciousness are dissociable in theory since dreaming dissociates them in practice. Dreams are exercises in "pure" phenomenality that show that "phenomenology is independent of cognitive access."[47] Dreams present dreamers with an experiential arena purged of cognitive control. Hence, they expose the shortcomings of two respected scholarly enterprises: cognitivist theories of consciousness that equate conscious experience with executive cognition,[48] and higher-order theories of mind that make the same conceptual move on philosophical rather than scientific terrain.[49] These theories assume that access consciousness is the plinth upon which all forms of conscious experience rests. However, dreaming shows us that that some conscious states depend on phenomenal consciousness alone.

Moral Status

We can now put the pieces of the argument together, which in a schematized form goes like this:

Premise 1	The foundation of moral status is phenomenal consciousness.
Premise 2	Dreams are phenomenally consciousness states.
Conclusion A	*Therefore, dreams confer moral status.*
Premise 3	Some animals dream.
Conclusion B	*Therefore, at least some animals have moral status.*

But moral status is a fuzzy philosophical concept whose practical consequences are far from settled.[50] In practice, what does it mean to attribute moral status to animals? Does it mean that we ought to be kind to them out of the goodness of our hearts? Does it mean that we are bound to take their interests into account when making decisions that affect them? Does it mean that we recognize them as having basic legal rights, such as the right to life and the right to bodily liberty? Does it mean that we cannot use them in scientific research, that we cannot display them at zoos and aquaria, that we cannot make them do physical or emotional labor for us? Does it mean that they cannot be domestic companions?

While I do not intend to tackle these huge ethical questions here, we must not miss the forest for the trees. The concept of moral status can do vital moral work for us even when we cannot fill in all the practical details. For example, the animal ethicist David DeGrazia explains that attributing moral status to animals is more than enough to condemn social institutions that oppress animals, such as industrial factory farming and invasive biomedical and behavioral research.[51] Moral status entitles animals to moral rights that protect them from having their interests infringed upon (if one is a consequentialist) or from having their dignity violated for the sake of comfort, expediency, or progress (if one is a deontologist). Institutions that oppress and exploit animals are moral calamities that no reasonable ethical framework, consequentialist or deontological, can justify. We do not need to have a full grasp of all the ethical, legal, and social implications of recognizing the moral status of animals to be sure that institutions such as factory farming and invasive scientific research will not clear the bar. The concept of moral status may be unnervingly slippery at the edges, but it is surprisingly steady at the core.

Recognizing the moral status of animals can help us navigate human-animal relations in a more ethical manner, bringing us closer to the goal of animal liberation. But I do not wish to simplify a complex topic. Moral status may be a powerful tool in the struggle for interspecies justice, but it is not a *nostrum* for all our societal problems. It is not a quick fix for the long and ghastly history of our treatment—or, rather, maltreatment—of animals. This concept may bring animals inside our moral universe, but it does not tell us the particular place they ought to occupy in it. Even if we attributed moral status to all animals, from the gnat to the blue whale, we would still have a lot of theoretical work to do. We would need to figure out what animals have what interests, what interests elicit what protections, and what protections trigger what consequences. As the animal ethicist Lori Gruen observes, "That non-human animals can make moral claims on us does not in itself indicate how such claims are to be assessed and conflicting claims adjudicated. Being morally considerable is like showing up on a moral radar screen—how strong the signal is or where it is located on the screen are separate questions."[52] But one thing is certain: we cannot get to the point of adjudicating conflicting claims until we grant animals access to the moral universe from which they are currently exiled.

Only moral status grants animals that access.

ETHICAL CODA

Contemporary denials of animal consciousness should frighten us because the distance between the repudiation of animal interiority and a total disregard for their well-being is infinitesimally small.[53] One of the main ethical challenges of our moment is to loosen the grip these denials have over our thinking so we can stop perceiving animals as mindless lumps of matter and start experiencing them as conscious beings who matter and for whom things matter, which is to say, as beings who—by virtue of their very existence—have value and imbue the world with value.

"Minding" animals is one way of making progress on this moral front.[54] I like this term because of its double meaning. It means seeing and treating animals as cognitive agents and caring about what happens to them, about the lives they lead and the conditions under which they live them. I also like it because its meanings interlock. Minding animals cognitively makes minding them ethically possible—or, at least, a whole lot easier. Projects that undermine the view that animals are mindless brutes can check the paroxysms of speciesist violence. By camouflaging itself as a legitimate consequence of our special standing as humans, this violence becomes all the more brutal as it reproduces itself without the slightest hesitation in so many areas of life. My contention is that we will not succeed at "minding" animals in this double sense unless we pay attention to all the facets of the animal mind, from the visible ones that animals display while awake to the hidden ones they nurse in the solitude of sleep.

EPILOGUE

Animal Subjects, World Builders

Dreaming, seemingly so inconsequential, has the curious attribute of leading us on into deeper and deeper philosophical issues.

—IAN HACKING[1]

There is a lot that we do not know about animals. Who are these nonhuman mortals with whom we share our brief existence? Who are they to us, and we to them? And given the many real forces that separate us (the gap of language, the problem of other minds, the perils of anthropomorphism, and so on), how should we understand the equally numerous and equally real counterforces that at the same time bind us together?

The Italian philosopher Paola Cavalieri calls this "the animal question."[2]

WHAT SEPARATES US

Our foray into the dream worlds of animals shows that animals are not deflated versions of us. They are not trapped in some preternatural state of arrested physical, psychological, evolutionary, ontological, or spiritual development. They have their own bodily schemas, psychic structures, and evolutionary histories; their own interests, aspirations, and motivations; their own ways of shaping and interpreting reality, of enduring and enjoying the sheer plenitude of the world. We may occasionally see aspects of our experience reflected in theirs, but they are not themselves reflections of us. They are not there to mirror or complement us. They do not exist for us or thanks to us. They exist to be who and what they are, and not who or what we wish them to be. They are, to borrow a term from the philosopher Tom Regan, "subjects of a life," which is to say, subjects of *their own lives*.

For us, this their-ness constitutes an inexorable limit. It means that our all attempts at understanding them will be beset by ambiguities that we cannot resolve, by questions that we cannot answer—or, at least, cannot answer *well*. And my own attempt to comprehend what happens when animals dream is no exception.

We are no closer now than at the start of this book to having a complete account of what other animals dream about. At most, we have a partial one. We know that they dream about pleasant things when their waking experience sparks their interest, curiosity, and joy. We know that they dream about the most horrendous things when trauma sinks its teeth into them, refusing to let go. And we know from our analysis of research on hippocampal replay that not all their dreams are palimpsests of past experience since some involve irreal phenomena without a referent in the real world. Even so, pressing questions remain unanswered, such as:

Exactly how high above the ground of experience can the minds of animals soar during sleep?
Can animals dream abstract thoughts?
Can they, in their dreams, solve problems that afflict them in their waking life?
Can they experience dream control, false awakenings, or sleep paralysis?
And just how bizarre, illogical, and surreal can their dream worlds be?
Do rats ever dream of being cats?
And do cats, in their dreams, ever find themselves at the other end of the chase?

The short answer is we have no idea. But if we grant that their dreams are not always faithful reruns of past events, we have to at least entertain the possibility that they may be as absurd, ingenious, and uncanny as ours—absurd, ingenious, and uncanny, of course, in a quintessentially nonhuman way.[3]

Aside from dream content, another issue we have left unresolved in this book is dream remembrance. As the philosopher José Miguel Guardia noted in 1892, "It would be curious to know whether animals remember their nocturnal hallucinations: that's a point neglected by authors who write for or against the spirit of animals."[4] More than a century later, neglected it remains—perhaps, I admit, for good reasons. It is hard to imagine how one would go about judging whether animals remember their dreams. Still, we know enough about the memory systems of other species to say that we cannot dismiss this possibility out of hand, even if we cannot say anything concrete about it just yet. Perhaps animals remember only some of their dreams and only for a brief period of time. Even so, this would mean that what happens to them in their dreams affects how they think, act, and live while awake. It would mean that their dream world bleeds into their waking world, making it tilt.

Dream remembrance would also mean that animals face the exceptional challenge of integrating their dream memories into their ongoing sense of self because, cognitively speaking, to dream and to remember a dream are two different things. As dream expert Ernest Hartmann observes:

> This basic function of dreaming occurs whether or not a dream is actually remembered. When the dream is remembered, it can have further functions in terms of revealing broader connections and possibilities useful in self-knowledge, life decisions, and new discovery.[5]

Antonio Damasio holds a similar view. In *Descartes' Error: Emotion, Reason, and the Human Brain,* he explains that our sense of who we are does not come, as Descartes believed, from above, from reason. It comes from below, from the slow and steady consolidation of emotionally colored memories, including dream memories. What would the discovery of dream remembrance in other animals mean, then, if not that we are not the only creatures who constantly spin, out of the strands of the past, what Damasio calls "an autobiographical sense of self"?

I recognize that these remarks bring us to the edge of a precipice. Here, we risk grappling with unanswerable questions, or with questions whose answers can only be tentative, imprecise, and full of disclaimers, if not outright speculative. But there is no way around this. On the one hand, this is what it means to study other animals. It means making peace with the trace of the indeterminate—and indetermin*able*—that mediates our relations to them. On the other, this is what it means to study dreams. As the philosopher of science Ian Hacking observes, dreams are just weird. As objects of study, they have what Hacking calls "the curious attribute" of drawing us in with their fantastical allure only to then lure us further and further away from our areas of intellectual comfort until we find ourselves in the bowels of a *terra incognita*, feeling stranded and disoriented. If our own dreams have this estranging and defamiliarizing effect on us, what should we expect from the dreams of the endless other beings that the unrelenting forces of evolution have also brought into this world?

We can turn melancholic and rue the fact that aspects of this world do not lend themselves to human mastery. Or we can bid welcome to the inherent opaqueness of the natural world and grow intellectually and spiritually as a result. Perhaps the experience of following animals into the vertiginous hinterland of dreams will pry us loose from our own ossified assumptions, especially our assumptions about the heights and depths their minds can reach and the immeasurable distances their souls can travel.

After disorientation, *re*orientation.

WHAT BINDS US TOGETHER

Animals have richly memoried, richly creative, richly embodied minds, and dreams give us a glimpse into this richness. More specifically, dreams make us realize that, like us, animals play an active role in the constitution of their own experience of the world. More than passively receiving experience ready-made, animals transform the chaotic flows of sense data that affect them into a single, meaningful, and coherent phenomenal world from within.

Nowadays neuroscientists and philosophers agree that all conscious experience is fundamentally creative. As the external world impresses itself on our senses through different forms of physical energy (such as light, temperature, pressure, chemical compounds), our mind-bodies transform these chaotic energy patterns into a unified phenomenal world equipped with spatiotemporal coordinates, stable percepts, emotional valence, social dynamics, and so on. At each and every moment of our existence, our mind-bodies are busy—and busily—weaving the disordered flows of data that bombard our senses into the "semiotically organized [. . .] field of meaning" that we call reality.[6] This creative impulse is always at work in our conscious life, independently of whether we are awake or dreaming.

What makes dreaming unique, however, is that it generates a field of meaning under the radical conditions of near-total sensorimotor blockade. Almost by definition, to dream is to perform a quasi-magical mental trick whereby we conjure into existence a subjective reality without the guiding hand of the external world. Indeed, if we had to name but one difference between our dreaming and waking lives, it would be just that—that dreaming depends less on what is "out there," while wakeful experience is in direct, nonstop conversation with it. As Allan Hobson says in *The Dream Drugstore: Chemically Altered States of Consciousness*, dreams are profoundly "auto-creative."[7] They are mental works of art that the mind creates for itself. This auto-creativity is spellbinding and bewildering since it leaves us with a string of questions for which we do not yet have solid answers, such as:

In the history of animal life, when did this auto-creativity first emerge and why?
By what long and winding paths did it find its way into so many branches of the evolutionary tree?
What spark or sparks does it produce inside the animal mind?
What types of subjective experience does it presuppose?
And what types of experience does it, in turn, enable?

While developing a full theory of animal subjectivity is beyond the scope of this book, I argue that this auto-creativity begins to limn the outline of such a theory by cueing us in to the extraordinary *world-building power* of animals—animals who, even in the oceanic calmness of sleep, give birth to enigmatic and imaginary worlds from the deepest depths of their being.[8]

Vanity lulls us into thinking that we alone possess this world-building power. As Friedrich Nietzsche argued in the late 1800s, our human pride makes us see "the entire universe as the infinitely fractured echo of one original sound—man; the entire universe as the infinitely multiplied copy of one original image—man."[9] But the universe is not merely an echo or a copy of us. "If we could communicate with the gnat," Nietzsche goes on to say, "we would learn that he likewise flies through the air with the same solemnity, that he feels the flying center of the universe within himself." In Nietzsche's account, all animals are "artistically creating subjects" who build up phenomenal realities tailored to their own existence. Even gnats build gnat-worlds by being gnats, by projecting *their* sound and *their* image onto the cosmos. Even their eyes, Nietzsche says, "glide over the surface of things and see 'forms.'"[10]

Darwin drew practically the same conclusion about the world-building power of other life forms a few years before Nietzsche when, quoting the German romantic Jean Paul Richter, he defined the dreams of animals in *The Descent of Man* as "involuntary acts of poetry."[11] Animals are accidental poets who invent "brilliant and novel results" by tirelessly combining and recombining the old and the new. In this spirit, I want us to sit with the idea that dreaming represents the art of subjective world formation, and that dreams are odes the animal mind sings to itself during sleep. By giving an audience to these odes, even when sung in inhuman tongues, we embark on the task of un-concealing the truth that our own arrogance has concealed from view: that, like us, animals are authors of their own experience and architects of their own realities; that, like us, they are builders of worlds, even after the Stygian currents of sleep have pulled them under and sent them flying through the looking glass.

NOTES

INTRODUCTION

1 Carson (1994), p. 25.
2 "Octopuses: Making Contact" aired October 2, 2019, on PBS.
3 Video available at: https://www.pbs.org/video/octopus-dreaming-trept6/.
4 Santayana (1940), p. 303.
5 The Roman philosopher Lucretius discusses animal dreams in *On the Nature of Things* (*De rerum natura*), which was written in the first century BCE. See Lucretius (1910), pp. 176–77.
6 Halton (1989), p. 9.
7 Aside from Manger and Siegel's article, the terms "dream" and "dreaming" are absent from practically all publications on animal sleep, even if they appear in media coverage of these publications in venues such as *National Geographic*, *The Independent*, and *BBC News*. An important exception is Malinowski, Scheel, and McCloskey (2021), which unfortunately is not discussed here as it was published after this book had already gone to press. The notion that at least some animals may dream does appear occasionally in publications on human dreaming, where experts seem to be more amenable to the idea. Jouvet ([1962]; [1979]; [2000]) and Hartmann (2001) are good examples. Nevertheless, most experts in the psychology and neuroscience of dreaming still hypostatize a wide and jagged fault line between the dreams of humans (which they insist are legitimate objects of empirical inquiry) and those of animals (about which, they tell us, we can say nothing concrete, sometimes not even that they exist).
8 Oneiric behaviors are a wide range of bodily movements that living organisms perform while asleep, especially during the phase of sleep typically associated with dreaming. These movements include rapid eye movements (REMs), sleep running, sleep fighting, sleep mumbling, etc.

9 "Mental replay" refers to patterns of brain activity that animals display during different phases of their sleep cycle and that seem to recapitulate waking behaviors.

10 When oneiric behaviors are observed in humans, they are interpreted as the behavioral correlates of a conscious experience, as outward expressions of an inward reality; but when they are observed in other animals, they are often presented as unconscious physiological events devoid of subjective significance. In a similar vein, when humans display specific patterns of neural activity during sleep, there is little doubt that they are "dreaming;" but when these patterns are spotted in other species, scientists switch discursive gears immediately, at which point the term "mental replay" takes over. These terms may sound synonymous, but they are not. The crucial difference is that dreams are *lived realities* that entail conscious awareness of some sort, whereas mental replay (as scientists define it) is a *cognitive process* that can unfold without conscious awareness.

11 Griffin (1998), p. 13. Experts in critical animal studies have traced the origins of mentophobia to various historical sources, such as the humanist prejudices of Western philosophy, the anthropocentric teachings of Judeo-Christian values, and the mechanistic spirit of the scientific revolution of the seventeenth century. Donald Griffin, for example, traces them to the theoretical commitments of behaviorist psychology (Griffin [1998]), while the French philosopher Vinciane Despret offers an alternative account, pointing to the late nineteenth century, when scientists began to forge their "professional identity" by differentiating themselves from other social actors who were also considered authorities on the subject of animals, such as "amateurs, hunters, breeders, trainers, caregivers, and naturalists" ([2016], p. 40). The abandonment of mentalistic concepts in the study of animal behavior was part of a larger strategy to crown science as the only reliable source of knowledge about animals, which could not happen without the disqualification of "a mode of thinking or knowing from which the scientific practice [. . .] tried to free itself, namely, that of the amateur" (p. 40).

12 Nonacademics are often surprised to learn that there are academics who openly defend the view that animals lack consciousness. The most famous of these is the philosopher Peter Carruthers, who has written numerous articles and books on the subject (see Carruthers [1989], [1998], [2008]). According to him, animals live in the same kind of mental darkness that we find ourselves in when we lose awareness of our surroundings while awake, such as when we "zone out" while driving for long periods of time.

He is not alone in this position. Joseph LeDoux, a renowned expert on the neuroscience of emotion, argues that we should stop attributing even basic emotions such as fear to animals since we cannot be certain that animals really experience these feelings (LeDoux [2013]). Hitting a similar note, the biologist Marian Dawkins encourages scientists to be "militantly agnostic" about the interior lives of animals ([2012], p. 177).

13 Curiously, Norman Malcolm's declaration that animals are thoughtless brutes is only one facet of his larger human-centric project. Another facet is his linguistic interpretation of dreams, which is developed in his book *Dreaming* (1959). Malcolm's position is not simply that only creatures endowed with language possess the subjective capacity to dream. In his view, the content of a dream is equal to its linguistic report. As the philosopher of science Ian Hacking has astutely observed, this makes it sound as if dreams are retroactively constituted by their own recall, and it leads to the ludicrous conclusion if we cannot recall a dream, then there must not have been a dream at all ([2004], p. 232). Malcolm explicitly weaponizes his position against animals, arguing not that the latter cannot report because they cannot dream, but rather that they cannot dream because they cannot report.

14 In general, external descriptions privilege objective states of affairs that are subject to quantitative measurement and intersubjective confirmation, whereas internal descriptions hinge upon subjective realities such as beliefs, intentions, and emotions that are hard to study using empirical methods. As the animal philosopher Elisa Aaltola explains, "Internal descriptions emphasize subjective experiences, whereas external descriptions underline mechanical explanations. Hence, the former explains intentional-seeming behaviour through the experiences and cognitive state of the animal, and the latter through matters such as mechanical instinct, behaviourism, and brain physiology" ([2010], p. 71).

15 Named after the psychologist C. Lloyd Morgan, Morgan's canon refers to the belief that complex explanations of animal behavior are less reliable than simple ones. Whenever possible, researchers must embrace low-level explanations that enlist only anatomical and physiological concepts. They can work their way up to high-level explanations that incorporate psychological and cognitive concepts only when low-level explanations prove deficient. One issue with this canon is that it is always possible, at least theoretically, to offer low-level explanations of even the most complex intentional and social behaviors. Furthermore, as the primatologist Frans

de Waal has observed, this canon can easily become a self-fulfilling prophecy. We demand low-level explanations of animal behavior because we assume that animals do not have complex minds, yet we believe that they do not have complex minds because we only find low-level explanations everywhere we look ([2016], pp. 42–45).

16 Unfortunately, many scientists who appeal to this philosophical argument are not well versed in the philosophical literature that surrounds it. As a result, they either take this argument to be more conclusive than it is or fail to notice that this view might commit them to a position that they do not actually wish to endorse. For instance, does the problem of other minds apply only to animals or to humans as well? If it applies to humans, does it vary across cultures, religions, and nationalities? And can it ever be solved? Can anybody ever know another person's mind? This seems unlikely. A strict application of this problem might lead directly to an extreme form of solipsism that most people would rightly reject. What weight, then, should the problem have in debates about animal consciousness?

17 Dream experts today interpret dreams that occur during REM sleep as triggered by sharp bursts of neuronal activity that begin in the pons (P), pass through the lateral geniculate nucleus (G), and culminate in the synthesis of a visual experience at the hands of the occipital lobe (O). First developed by Allan Hobson and Robert McCarley in the late 1970s, this "activation-synthesis hypothesis" is a standard lens for the neuroscientific study of dreams (see Hobson and McCarley [1977]). As I explain in chapter 1, versions of the human PGO wave have been found in other species, such as zebrafish.

18 This is not to say that verbal reports are not used in dream science, but it is to say that much of the scientific research on dreaming published today is about the behavioral and neuroscientific dimensions of dreams. Moreover, excitement about verbal reports has waned as researchers have come to accept that said reports suffer from a host of drawbacks related to the fidelity of memory-recall.

19 We will never know with absolute certainty whether other animals dream because we have no direct access to their inner lives. But science does not trade in absolutes. By design, its most powerful claims are probabilistic and defeasible judgments whose epistemic value comes from having more or less support, not from being true in an absolute sense.

20 In positioning myself between science and philosophy, I am inspired by Poovey's (1998) account of the social construction of scientific facts.

21 In writing this book, I have relied on the insights of other scholars who have come to the defense of animal dreaming, such as Gay Luce, Michel Jouvet, Ernest Hartmann, Kenway Louie, Matthew Wilson, Paul Manger, Jerome Siegel, Marc Bekoff, and Boris Cyrulnik. Here, I build upon their work while also going beyond it in three important ways. While they all mention evidence of animal dreaming, no synthesis has been attempted of the kind I present in chapter 1. Moreover, in this book I tarry with the philosophical implications of animal dreams more systematically than any of these thinkers do. Finally, there is the question of scope. This introduction is already longer than what all these thinkers, with the exception of Jouvet, have published on the subject.

CHAPTER 1

1 Darwin (1891), p. 169.
2 Lindsay (1879), p. 94.
3 Lindsay (1879), p. 95.
4 Morse and Danahay (2017).
5 Kant believed animals have "reproductive" imagination, which recalls past events. He denied they have "productive" imagination, which is what most people think of when they use this term in everyday speech (Fisher [2017]). Productive imagination is about creating new things, even things that cannot be directly experienced. For a discussion of animal imagination, see chapter 3.
6 Romanes claimed that imagination comes in four degrees. Imagination in the first degree occurs when perception of an object causes an animal to recall properties of the object that are not given through the act of perception (for example, when I see an orange from far away and I recall the smell). Imagination in the second degree occurs when an animal mentally visualizes an object that is not present in their environment because another object that is present reminds them of it (such as when I see water and am reminded of wine, so I visualize a glass of wine). Imagination in the third degree happens when we envision an object spontaneously and at will, without a cue from our surroundings. Finally, imagination in the fourth degree involves "intentionally forming mind-pictures with the sole purpose of obtaining new ideal combinations" ([1883], p. 144). Romanes did not believe that animals have imagination in the fourth degree, which is

"distinctive of Man" (p. 144), but he was sure that they have imagination in the first three degrees. Dreams in particular, he argued, belong to the third degree since they involve visualizing objects without external cues (p. 148). A similar argument was made three centuries earlier by the French moralist Michel de Montaigne, who wrote: "Even brute beasts are subject to the force of imagination as well as we; witness dogs, who die of grief for the loss of their masters; and bark and tremble and start in their sleep; so horses will kick and whinny in their sleep" (Montaigne [1877]).

7 Romanes (1883), p. 148.

8 Romanes (1883), p. 148.

9 Also mentioned were Jean-Charles Houzeau, Robert Macnish, Johann Bechstein, Thomas Jerdon, and Georges-Louis Leclerc, Comte de Buffon.

10 Philosophers were also fascinated by this topic. The Spanish philosopher José Miguel Guardia talked about the dreams of animals in an 1892 article he published in the journal *Philosophical Review of France and Abroad* [*Revue Philosophique de la France et de l'Étranger*]. This piece influenced the father of psychoanalysis, Sigmund Freud, who would later mention the dreams of animals in his seminal *On the Interpretation of Dreams*. In his review of De Sanctis's work, Binet also references the work of the American philosopher Mary Whiton Calkins, who wrote an important essay on the statistics of dreams and went on to become the first woman president of both the American Psychological Association and the American Philosophical Association, in 1905 and 1918 respectively.

11 Behaviorism, which reigned supreme in psychology for the first half of the twentieth century, rejected the intuitionist methods of nineteenth century psychology and maintained that in order for psychology to become a science it would have to limit itself to the study of publicly observable facts, i.e., behavior. This meant eschewing mentalistic concepts such as "mind," "idea," "symbol," "schema," "thinking," "feeling," and "representation." The cognitive revolution of the 1950s and 1960s resisted this approach and demanded that psychology return to its original mission of mapping the internal structure of the mind, which meant reintroducing into the discourse of psychology all the concepts that behaviorists vilified. By the 1970s and 1980s, psychologists had cut most of their ties to behaviorism and were once more debating mental states. Yet, even as they moved away from behaviorist principles in the study of human behavior, many psychologists continued to openly defend their use in the study of animal behavior, thereby ceding terrain to behav-

iorism in fields such as evolutionary biology, zoology, and ethology. Readers interested in the cognitive revolution should consult Gardner (1987) for a historical survey, and Baars (1986) for a philosophical defense.

12 de Waal (2016).

13 Foulkes's (1990) position represents the mainstream. Because dreams are acquired gradually during human childhood, usually in sync with symbolic abilities, it is often assumed that they depend on those abilities.

14 Dumpert (2019).

15 The songs of zebra finches are complex musical achievements composed of notes, which make up syllables, which in turn make up motifs. These songs are learned rather than innate (Derégnaucourt and Gahr [2013]).

16 Dave and Margoliash (2000), p. 815.

17 Dave and Margoliash (2000), p. 812.

18 When I say that replay has no accompanying phenomenology, what I have in mind is that, in the words of the philosopher Thomas Nagel, there is nothing "it is like" for the finches to undergo replay. Phenomenal consciousness, as Nagel defines it, is associated with qualitative experiences such as seeing, smelling, tasting, and being in pain (Nagel [1974]). Ned Block also includes in this category subjective states that do not hinge on sensing external stimuli, such as private thoughts, desire, emotions, feelings, and internal sensations (Block [1995]).

19 Neurobiology recognizes the importance of the temporality of neural events. Thompson (2007) borrows Varela's (1999) distinction between the "1/10 scale" the "1 scale" of neural activity to argue that events in the 1/10 scale (between 10 and 100 milliseconds) may be too quick to have a phenomenological correlate, but things unfolding at the 1 scale (between 250 milliseconds to several seconds) can constitute what is "happening now" for a subject. "This neurodynamical now," Thompson writes, "is the neural basis for the present cognitive moment" (p. 334).

20 "CA" stands for *cornu Ammon*, which is Latin for "Ammon's horn," whose shape resembles that of the hippocampus.

21 Dehaene (2014) discusses some of this research. He notes that if one tricks rats into thinking that they are located in a different position than where they think they are (say, by painting the walls in their enclosures or changing the soil), hippocampal cells "vacillate" between both interpretations until settling on one, depending on how successfully the illusion is carried out (p. 207).

22 Louie and Wilson (2001), p. 154.
23 Louie and Wilson (2001), p. 146.
24 Louie and Wilson (2001), p. 149. They point out that the slower progression of hippocampal activity during REM sleep could be due to temperature differences between the waking and the sleeping state. "The frequency of the theta rhythm is sensitive to brain temperature and brain temperature is typically lower during sleep, suggesting that the neural process underlying REM reactivation may be similarly slowed" (p. 154).
25 Bendor and Wilson (2012) have shown that it is possible to alter the content of a rat's dreams via replay experiments. By exposing animals to different stimuli while they sleep, they can alter brain activation patterns in ways that are likely to bring about new dream experiences.
26 Louie and Wilson (2001), p. 149.
27 Louie and Wilson (2001), p. 151.
28 Louie and Wilson (2001), p. 153.
29 Brereton (2000) observes that hippocampal activity during REM sleep matches hippocampal activity during the waking state but not during non-REM sleep. Citing Rotenberg (1993), he explains that there is clinical evidence "that large, slow, theta rhythms are present in the hippocampi of rats, rabbits, and cats in two different metabolic states: waking search and survival activities, and REM sleep" (p. 387).
30 The term "simulation" is crucial and appears in human dream research often, especially in Antti Revonsuo's theory of dreaming, which holds that dreams are internally generated simulations of reality (Revonsuo [2000], [2005]).
31 Leung et al. (2019), p. 201. The authors concluded that these states are distinct because they are affected differently by sleep deprivation. When zebrafish are sleep deprived, they need "sleep rebound" for SBS, but not PWS. This is consistent with research about the effects of sleep deprivation on mammals. Mammals, too, need sleep rebound for non-REM sleep, but not for REM sleep. One difference between PWS in fish and REM sleep in mammals is the absence, in the former, of REMs (p. 201). We must be careful not to equate the absence of REMs, however, with the absence of dream experience since behavioral markers of dream experiences may vary across species.
32 Leung et al. (2019), p. 203. MCH neurons are periventricular ependymal cells that express MCH (melanin- concentrating hormone). Ablation experiments have established that when MCH2 neurons are damaged, zebrafish

exhibit disrupted sleep patterns, including an overall decrease in the amount of sleep during the night. "These results show that MCH signaling has an important role in activating the PWS signature, and in regulating the amount of sleep, in zebrafish" (p. 203).

33 Solms (2021), p. 26ff.

34 Leung and his team claim to only study these signatures "in an agnostic fashion" ([2019], p. 198), which is to say, without taking a position about the attending phenomenology. They claim that the "agnostic identification" of the underlying neural factors associated with SBS and PSW helps us understand the deep evolutionary origins of our modern biphasic sleep cycle, which evolved before "the radiation of amniotes" (mammals, birds, and reptiles, p. 203). It is this bizarre agnosticism that, I argue, continually gets in the way of the articulation of a theory of animal dreams. As in the case of Dave and Margoliash above, this position prevents Leung's team from recognizing the possible bearing of their findings on the question of animal mentation during sleep.

35 Video available at www.youtube.com/watch?v=wI8XgJ3JebE.

36 See Introduction, endnote 15.

37 Preston (2019), p. 1.

38 Two other facts are relevant. One is that Heidi is not the only octopus who has been filmed making chromatophoric displays during sleep (Starr [2019]). Another is that octopuses are not the only species who make synchronically controlled and diachronically coherent displays. Platypuses do too. Siegel et al. ([1999], p. 392) have shown that when platypuses go into REM sleep, they often make the same mastication movement that they make when eating freshwater crustaceans, one of their favorite foods.

39 See Chase and Morales (1990).

40 I borrow the distinction between response and reaction from Derrida (2002), who uses it to distinguish intentional action (action that makes sense only in relation to the interests, goals, and desires of an organism) from unconscious reaction (mechanical behaviors, such as reflex reactions, that can be understood without reference to phenomenological or psychological concepts).

41 Certain sleep behaviors, such as REMs, are widely but not universally recognized as behavioral indices of dream phenomenology. One critic of this position is Blumberg (2010), whose "ontogenetic hypothesis" stakes itself on the claim that there is nothing special about REMs since these can occur

even in "transected" animals (animals whose cortex and brainstem has been disconnected).

42 Frank et al. (2012), p. 5. Godfrey-Smith writes, "Cuttlefish appear to have a form of rapid eye movement (REM) sleep, like the sleep in which we dream" ([2017], p. 73).

43 Frank et al. (2012), p. 2.

44 This echoes similar findings by Duntley, Uhles, and Feren (2002) and Duntley ([2003], [2004]).

45 Frank et al. (2012), p. 2.

46 Louie and Wilson explicitly say that the appearance of the same neural pattern as RUN during REM sleep could not be due to chance ([2001], p. 147).

47 Frank et al. (2012), p. 5.

48 Frank et al. (2012), p. 5.

49 On humans, see MacWilliam (1923), Aserzinsky and Kleitman (1953), Snyder et al. (1964), Nowlin et al. (1965), and Somers et al. (1993). On cats, see Baccelli (1969), Baust and Bohnert (1969), Baust, Holzbach, and Zechlin (1972), and Rowe et al. (1999). On dogs, see Kirby and Verrier (1989) and Dickerson et al. (1993). On rats, see Sei and Morita (1996).

50 Lacrampe (2002).

51 Leung et al. (2019).

52 Rowe et al. (1999), p. 845.

53 Corner (2013) argues that the displays of cuttlefish during REM sleep are analogous to their waking displays.

54 Frank et al. (2012), p. 5.

55 Frank et al. (2012), p. 5. They found that juvenile cuttlefish do not present REM sleep. Only adult cuttlefish do. They impute this finding to "differences in neural maturation" (p. 6).

56 This study, "Comprehensive Nighttime Activity Budgets of Captive Chimpanzees (*Pan troglodytes*)," was Mukobi's master's thesis at Central Washington University.

57 Mukobi (1995), p. 59.

58 Of course, a few hand twitches do not an ASL sign make, but Mukobi reminds us that when humans talk in their sleep, they often mumble and utter incomprehensible things. "Therefore, it should not be expected that a sign occurring during sleep be as precise as when it is signed during wakefulness" ([1995], p. 58).

59 Mukobi (1995), pp. 47–48.

60 When I first read about Washoe's sleep sign for coffee, I wondered why a chimpanzee would know what coffee is. It turns out that the chimpanzees at the Chimpanzee and Human Communication Institute occasionally drank coffee. And to get it, they had to ask for it. "We had a coffee maker in the kitchen and there were big windows between the kitchen and some of the indoor chimp rooms so they would watch us make food, pour coffee, chat with each other, and anything else we'd do in the kitchen. If they wanted something extra, they would let us know and this included coffee once in a while. It's not like they drank it every day but occasionally, they would express interest and someone would make them a cup to take a sip (cooled off, of course)" (Mukobi, personal email). Allen and Beatrix Gardner, who cross-fostered several of the chimpanzees in Mukobi's study, taught them this concept from a young age. See Van Cantfort, Gardner, & Gardner (1989).

61 Mukobi (1995), p. 59.

62 Mukobi, personal email.

63 She cites Karacan, Salis, and Williams's 1973 study of sleep talking in humans, which "concluded that talking in one's sleep may be another indication of dreaming" ([1995], p. 7).

64 Mukobi (1995), p. 59.

65 Mukobi (1995), p. 56. I will have more to say about animal nightmares in chapter 2.

66 Ridley contends that our dreams are more "vivid" than those of animals ([2003], p. 16), while Hartmann says that the latter are also "less complex and less metaphoric" than ours ([2001], p. 211).

67 Jouvet (2000), p. 2.

68 Jouvet called it paradoxical sleep because the fast cortical EEG activity emanating from the PGO circuit during REM sleep was essentially identical to that of the waking state, yet people in this phase of sleep didn't behave as if they were awake. They were mostly immobile, with the exception of REMs. Jouvet insisted that the difference between paradoxical and non-paradoxical sleep was supported by electrophysiological (Jouvet [1962]), ontogenetic (Valatx, Jouvet and Jouvet [1964]), and phylogenetic (Klein [1963]) evidence. See Jouvet (1965b).

69 Quoted in Haselswerdt (2019), p. 3.

70 Jouvet (2000), p. 43.

71 In choosing cats, Jouvet followed the lead of the German scientist Richard Klaue, whose research on cats in the 1930s led to the identification of REM sleep as a distinct phase of sleep.

72 Jouvet (1965a) emphasizes that although the cats were standing up and moving around, they were sleeping. This was evinced by the fact that "the nictitating membranes are relaxed and may cover the pupils." One can watch the original video recordings of these cats at: https://www.youtube.com/watch?v=Js50Orx94iM.

73 Jouvet writes, "Thus, the hypothesis that a cat dreams of actions characteristic of its own species (lying in wait, attack, rage, fight, flight, pursuit) during its paradoxical sleep is quite plausible" ([2000], p. 92).

74 Brereton (2000), p. 393.

75 Pagel and Kirshtein (2017), p. 37.

76 An example of unreliable sleep behaviors is "myoclonic twitching," which refers to a spasmodic behavior observed during sleep in infants and embryos. This term, however, is sometimes used quite broadly, to the point that it includes the tossing and turning that occurs naturally during deep sleep as well as the wanderings of somnambulists, which typically occur during non-REM sleep and are rarely accompanied by dreams.

77 In Manger and Siegel's account, monotremes are unlikely to dream because there is no communication between brainstem and cortex during sleep. The same is true for cetaceans but for different reasons. Cetaceans are unlikely to dream because, when cetaceans sleep, half of their brain stays awake, and such unihemispherical sleep may be incompatible with dreaming, especially if the waking hemisphere is actively engaged with the external world. Finally, pinnipeds are similar to cetaceans, except that they switch between bilateral and unihemispherical sleep at different points of the year. African elephants, Arabian oryxes, rock hyraxes, and manatees are borderline cases because their sleep cycle is so different from that of most other mammals that we simply do not know whether the difference between REM and non-REM sleep even applies. While I find Manger and Siegel's approach to the question intriguing, I believe there is room to think that even some of their exceptions might not be exceptions after all. In chapter 2, for example, I discuss the nightmares of young African elephants.

78 Manger and Jerome (2020), p. 4. The problem is that dolphins are unihemispherical sleepers, and some experts believe that dreams are incompatible with unihemispherical sleep itself. This is the view of Mann (2018) and Jouvet

([2000], pp. 20–21). Yet, as Frank observes, not everyone agrees with this conclusion since it is not certain that cetaceans *cannot* go into REM sleep ([1999], p. 28). Zepelin (1994) cites several studies indicating that cetaceans go into REM sleep during periods of rest and display REMs, phasic motor activity, and even penile erections (which are very common in human males during REM sleep). Similarly, Shurley et al. (1969) reports REMs and motor atonia in the pilot whale, which is the second largest species of dolphin.

79 Burgin et al. (2018).
80 Jouvet (2000), p. 123.
81 Jouvet (2000), p. 123.
82 Siegel et al. (1998).
83 Nicol et al. (2000).
84 Edgar, Dement, and Fuller (1993).
85 Lyamin et al. (2002).
86 Dave and Margoliash (2000).
87 Lesku et al. (2011).
88 Stahel, Megirian, & Nicol (1984).
89 Berger & Walker (1972); Dewasmes et al. (1985).
90 Van Twyver & Allison (1972); Walker & Berger (1972); Graf, Heller & Rautenberg (1981); Graf, Heller & Sakaguchi (1983).
91 Lacrampe (2002), p. 67.
92 Lacrampe (2002), p. 67.
93 Underwood (2016).
94 Frank (1999), p. 24.
95 Frank (1999), p. 24.
96 Frank (1999), p. 24.
97 Jouvet claims "no one has yet been able to record unequivocally a state similar to paradoxical sleep in fish, amphibians, or reptiles (except perhaps in the crocodile)" ([2000], p. 55).
98 Lacrampe (2002), p. 51.
99 Duntley, Uhles & Feren (2002), Duntley (2003), Duntley (2004), and Frank et al. (2012).
100 Godfrey-Smith (2017), p. 1.
101 Jouvet explicitly excludes fish, amphibians, and reptiles, arguing that that "no one has yet been able to record unequivocally a state similar to paradoxical sleep in fish, amphibians, or reptiles (except perhaps in the crocodile)" ([2000], p. 55).

102 See Karmanova and Lazarev (1979) and Karmanova (1982).
103 Corner and van der Togt argue that active sleep evolved "in an ancestral reptile prior to divergence of the avian and mammalian lineages" ([2012], p. 27), but research on cuttlefish suggests that active sleep may be an example of parallel evolution.
104 Freiberg (2020). Jouvet shares this view, writing: "Dreams are difficult to recognize in a bacterium, an oyster, or a mosquito" ([2000], p. 55).
105 Hartmann concedes that animals dream and specifies that their dreams are likely to involve "a mix of sensory modalities different from ours" ([2001], p. 211).
106 Research on the dreams of humans who are blind supports this principle. See Hartmann ([2001], pp. 211ff).
107 Uexküll (2013).
108 Wittgenstein (1958), p. 223.
109 I borrow this reference from Pagel and Kirshtein, who write: "Dreams as experienced by animals may very well be like the thoughts of Wittgenstein's lion [. . .] They are likely to be radically unlike the dreams experienced by humans" ([2017], p. 40).
110 Romanes (1883), p. 149.
111 Bachelard (1963), p. 20.

CHAPTER 2

1 Steiner (1983), p. 6.
2 Andrews (2014) divides contemporary arguments for animal consciousness into four groups: (1) *representationalist* arguments that look at whether animals possess representational mental states, (2) *NCC (Neural Correlates of Consciousness)* arguments that isolate structural and functional similarities between the central nervous systems of human and nonhuman animals, (3) *self-consciousness* arguments that turn on evidence of self-recognition and mental monitoring in animals, and (4) *non-inferentialist* arguments that deny we need reasons to infer that animals are conscious because we immediately grasp animals as conscious in our interactions with them. Not included in Andrews's taxonomy are evolutionary theories, such as Mallat and Feinberg's (2016) theory that "primary consciousness" emerged during the Cambrian explosion. It is

worth noting that none of these approaches give much weight to what the minds of animals do while asleep.

3 I define dreaming as a sufficient but not necessary condition for consciousness because there are individuals who are conscious but who do not dream.

4 Rock (2004), p. 186. Some researchers turn this causal relationship around. Brereton (2000) interprets dreaming as a pre-adaptation that set the stage for the emergence of human consciousness.

5 Searle (1998), p. 1936.

6 Churchland (1995), p. 214. This view is built into the blueprint of the eliminativist program of contemporary neuroscience. Eliminativists embrace what Alva Noë dubs "the foundation argument" ([2009], p. 173), which holds that the only thing we need to achieve conscious awareness is a functional brain. In this view, a functional brain is "the engine of reason," "the seat of the soul" (see Churchland [1995]). In support of this view, eliminativists point to dreaming. When we dream, they say, we are conscious even though the neurochemistry of sleep immobilizes the body and severs our sensory connection to the external world. In this state, we sustain conscious awareness with nothing more than a functional brain. I reject this argument, which I interpret as the result of "neuro-nihilism" (Thompson [2015]). Research into the 4Es (embodied, embedded, extended, enactive) of cognition tells us that conscious experience requires three ingredients: a brain, a body, and a world. Without any one of these elements, conscious awareness cannot manifest. Still, the key point here is that idea that dreaming is a sufficient condition for consciousness is shared by eliminativists (Churchland [1995]) and anti-eliminativists (Noë [2009]).

7 The German phenomenologist Edmund Husserl held this view (Kockelmans [1994], p. 167), which was later popularized by Malcolm ([1956], [1959]).

8 Thompson (2015), p. 14.

9 Thompson (2015), p. 16.

10 Windt and Metzinger (2007), p. 194.

11 Of course, the dreamscape is not a "here and now" in the epistemological sense that it corresponds to mind-independent states of affairs unfolding in the external world. It is a "here and now" in the *phenomenological* sense that I experience it as real. For my purposes, it suffices to emphasize that the dreamscape is not *the* "here and now," but *a* "here and now."

12 Windt and Metzinger (2007), p. 194.

13 Windt and Metzinger (2007), p. 195.
14 Miller (1962), p. 40.
15 Dehaene (2014), p. 23.
16 Many taxonomies of consciousness have gained notoriety in a wide cross section of disciplines in the twentieth and twenty-first centuries. Some of these are relatively recent, while others have been "recovered" from earlier historical periods and made relevant once more. Consider the following examples: Sigmund Freud, the founder of psychoanalysis, distinguishes between "conscious," "preconscious," and "unconscious" mental states; the German philosopher Edmund Husserl differentiates between "thetic" and "pre-thetic" conscious modalities; the American philosopher Ned Block differentiates between "access" and "phenomenal" consciousness; the Portuguese-born neuroscientist Antonio Damasio oscillates between a dyadic and a triadic framework (he sometimes divides consciousness into "core" and "autobiographical," and sometimes into "proto-self," "nuclear self," and "autobiographical self"); working at the intersection of physics and neuroscience, the physicist John Taylor cuts it into "passive," "active," "self-aware," and "emotional" types; Richard Schmidt, a linguist, opts for "awareness," "intention," and "knowledge"; and Arthur Deikman, a psychiatrist, gives us "thinking," "feeling," "functional capacity," and "the observing center." Of course, this is not an exhaustive list, as plenty other stories have been told about what goes into the making of conscious life. One thing that bears mention is that there is little terminological consistency across these taxonomies. Some people use different terms for what seem to be similar strata of consciousness (such as "feeling" and "emotion"), while others use the same term for rather dissimilar strata. Zahavi (2014), for instance, reminds us that there are a thousand and one possible meanings of "the self."
17 Notice that my model of consciousness does not capture different *states* of consciousness, like Thompson (2015). It captures different forms that consciousness can take. This means that animals can have one, two, or all three forms while awake or while in a dream. My focus here will be on how these forms of conscious awareness congeal through the dream state, leaving aside the question of other states of consciousness such as the waking state, dreamless sleep, or what Thompson calls "pure awareness."
18 Zahavi (2014), p. 14.
19 Zahavi (2014), p. 14.
20 DeGrazia (2009), p. 201.

21 Some experts on animal cognition, such as Carruthers (2008), assert that animals cannot be subjectively conscious because they do not have metacognitive access to their own mental states. I, however, exclude metacognitive functions from my definition of subjective consciousness since I believe it is possible for an organism to be self-aware without metacognitive access to its own mental states. While some have noted that animals can reflect upon their own mental states (Andrews [2014]), others have pointed to evidence of other capacities that indicate subjective awareness, especially self-recognition, empathy, and deception (Gallup [1977]; Bekoff [2003]; de Waal [2016]).

22 Support for the idea that dreams are coherent spatiotemporal manifolds first came from the discovery that damage to the parietal lobe, a brain region in charge of producing mental images and spatial representations, stops dreams from occurring altogether. Neuroscientists have interpreted this to mean that dreams are not, to borrow William James's famous phrase, a "great, blooming, buzzling confusion." On the contrary, they are "the experience of an unfolding space-time continuum within which we move, feel, act, with the self at the center of experience" (Bogzaran and Deslauriers [2012], p. 47).

23 The notion of panoptical vision appears in Brereton ([2000], p. 393).

24 Windt (2010), p. 304 (emphasis added).

25 Windt (2010), p. 297. This position is also defended by Bogzaran and Deslauriers ([2012], p. 79).

26 Thompson (2015), p. 123.

27 Thompson (2015), p. 124.

28 Thompson (2015), p. 127. While I defend an egocentric interpretation of dreaming on phenomenological grounds, this interpretation is also backed by empirical data. Neurofunctionalist theories, such as Llinás and Paré (1999), trace waking and dreaming to the same neural mechanisms. Meanwhile, evolutionary theories, such as Revonsuo (2000), posit parallels between waking and dreaming at the level of evolutionary function.

29 Windt ([2010], [2015]) argues that the dream ego is not only phenomenally but also *functionally* embodied insofar as dream content can be affected by what happens to the dreamer's body during sleep. Since specific sensory inputs to the body can yield determinate dream outputs, dreaming "remains connected in interesting and systematic ways to the sleeping body" (Windt [2015], p. xxiii).

30 Sartre (2004), p. 166.

31 For Sartre, "all phenomena of attention have a motor basis" ([2004], p. 43). All attention is embodied in the sense that it depends on sensorimotor knowledge. Because dreams are special cases of conscious attention, they find their basis in motor movement and thus presuppose the possession of a body schema. Sartre does not deny that the body schema can be malleable in dreams. In a dream, I can easily have two heads, one cyclopean eye, or a thousand tentacles, but what I cannot have is no body schema at all.

32 Revonsuo (2005), p. 207.

33 Brereton (2000), p. 385.

34 Sartre (2004) borrows this concept from the German-American psychologist Kurt Lewin. This concept plays a key role in Sartre's discussion of "lived space" in his 1943 book *Being and Nothingness.*

35 Windt (2010) makes this claim. Similarly, Thompson (2015) argues that even the dreams of children and the dreamlike sequences we experience during hypnagogic sleep, which are frequently paraded as examples of egoless dreams, are organized around, and grounded in, an ego. Children's dreams are egocentrically arranged even if children do not have the capacity to attend to, and report, their subjective organization (p. 131ff). Similarly, even if hypnagogic images have "relaxed ego boundaries," they are "hardly free from the identifying and appropriating functions of the 'I-Me-Mine'" (p. 126). Sartre (2004) disagrees on this last point since he sees hypnagogic images as "dreams without the Me" (p. 166).

36 Godfrey-Smith (2016), p. 12.

37 Crick and Mitchison (1983).

38 In chapter 2 of *On the Interpretation of Dreams,* Freud recognizes that dream interpretation has a long cultural and philosophical history but argues that earlier approaches—such as the "symbolic method" of the Biblical Joseph or the "cypher method" of Artemidoros in the second century CE—were unscientific and unpsychological. As such, they suffered from the same limitation: they were utterly worthless. They were on the right track in assuming that dreams are meaningful, but they misunderstood their meaningfulness. The symbolic method assumed that their meaning lies in their power to foretell the future (as with Joseph), while the cypher method assumed that their meaning can be easily decrypted "according to an established key" that the interpreter can apply mechanically in the same way to all dreams.

39 Conn (1974), p. 711. For a more detailed account of the fall of psychoanalysis after the Second World War, see Hale (1995).

40 Damage to the vmPFC can bring dreaming to a halt (Rock [2004], pp. 46, 104).

41 Solms (2021), p. 27.

42 Brereton ([2000], p. 391). Limbic structures involved in dreaming include the fusiform gyrus (face recognition), the thalamus (body image), the cerebellar vermis (spatial and bodily movement), and the right parietal operculum (spatial location). Neuroimaging studies by Baird, Mota-Rolim, and Dresler (2019) show increased cerebral blood flow to these structures during REM sleep.

43 Rock (2004), p. 122.

44 Bogzaran and Deslauriers (2012), p. 48.

45 Bogzaran and Deslauriers (2012), p. 63. The link between dreaming and emotions is so tight that, according to Hartmann, dreams have a "quasi-therapeutic function" as they provide a mental framework for processing trauma ([1995], p. 180). This view has been developed more recently by Mark Solms in his book *The Hidden Spring: A Journey to the Source of Consciousness*, where he presents a systematic theory of consciousness rooted in emotions, feelings, and affects.

46 Damasio (1999), p. 100.

47 There was no such correspondence between the neural pattern connected to exploring the baited arm and either (a) the pattern exhibited during sleep before exposure to the maze or (b) the one associated with physical exploration of the uncued arm.

48 "Together," Ólafsdóttir and her co-authors write, "these findings indicate that biased pre-activation of future experiences [in rats] is instantiated at the point when an environment becomes motivationally relevant" ([2015], p. 10).

49 Much research on hippocampal replay associates pre-play with non-REM sleep. In itself, this does not automatically render pre-play incompatible with dreaming since dreams also occur during non-REM sleep, even if less frequently. Moreover, a subsequent study by Ólafsdóttir, Bush, and Barry (2018) established that the relevant spikes occur also during REM sleep. Citing Louie and Wilson (2001), they point out that during REM sleep these spikes "progress at a more natural speed" (p. R38) that is comparable to the spike events recorded during waking experience. When they occur

during non-REM sleep, they are about twenty times faster. Further evidence for a link between REM sleep, emotion, and memory comes from Boyce et al.'s (2016) study of theta rhythms. When the theta rhythm is inhibited while rats are in REM sleep, they are less likely to consolidate their memory of previous events during sleep, including memories of physical objects and unpleasant experiences (p. 815). In this context, we must also look at Karlsson and Frank's (2009) findings that patterns of hippocampal replay in rats "were present across time scales similar to those seen during experience" (p. 7). This claim is supported by Gelbard-Sagiv et al. (2008) and Pastalkova et al. (2008) and mentioned in passing by Knierim ([2009], p. 422).

50 Voltaire (1824), p. 118.

51 Berntsen and Jacobsen ([2008], p. 1093), emphasis added. Ólafsdóttir, Bush, and Barry (2018) explicitly state that during pre-play, rats are "planning forthcoming actions" (p. R43). I will have more to say about imagination in chapter 3.

52 Freud (1938), p. 215.

53 Freud (1938), p. 215.

54 The researchers exploited the fact that rats are highly empathic animals who, as Hernandez-Lallement et al. (2020) demonstrates, suffer immensely when they see *others* in pain.

55 Yu et al. (2015), p. 11.

56 Yu et al. (2015), p. 9. In chapter 1 we saw that many experts in animal sleep regularly turn away from the phenomenological dimensions of their findings. The same happens in this case. The authors write, "We cannot unequivocally state that the rats were actually experiencing a traumatic memory before being startled awake" ([2015], p. 10). This is an especially puzzling claim since the term "nightmare" appears in the title of their publication.

57 Van der Kolk (2015), p. 84.

58 Yu et al. (2015), p. 9. The authors replicated their findings in a follow-up study the following year (in 2016), this time looking at the neurochemistry of startled awakenings. One key finding of the follow-up study was that the traumatized rats experienced startled awakening at a moment in their sleep cycle when they had low levels of orexin, a neuropeptide that helps organisms naturally wake up from sleep. That the rats woke up in a panic despite having low levels of orexin suggests that their awakening was categorically different from the process of normal awakening in the absence of trauma (Yu et al. [2016]). It was a psychological, rather than a

physiological, awakening. Their minds, heavy with trauma, plucked their bodies from sleep.

59 Rats displayed freezing behavior and startled awakening for up to twenty-one days after the onset of trauma. We can only guess for how much longer these behaviors would have continued since the protocol culminated in the rats being killed so their brains could be isolated and analyzed.

60 Berardi et al. (2014), p. 8.

61 Campbell and Germain (2016); Vanderheyden et al. ([2015], pp. 2343–45).

62 Kirmayer (2009), p. 5.

63 Bradshaw (2009); Balcombe ([2010], p. 59); Cavalieri ([2012], p. 130).

64 Peña-Guzmán (2018), p. 16.

65 Masson (2009), p. 45. For a discussion of this orphanage, see Siebert (2011).

66 King (2011), p. 77.

67 Bradshaw (2009), p. 139.

68 Kingdom (2017). Reports of elephant nightmares highlight the dangers of false negatives. In their study of mammalian dreaming (discussed in chapter 1), Manger and Siegel (2020) argue that elephants are not good candidates for dreaming because of their peculiar sleep cycle. This is an example of how we may be led astray by using human sleep as the default reference point in our thinking about the dreams of animals.

69 Morin (2015); Bender (2016).

70 The haunting video of Michael's signing is available online at: www.youtube.com/watch?v=DXKsPqQ0Ycc.

71 Kelly (2018).

72 Botero (2020), p. 4.

73 Botero (2020), p. 4.

74 Maternal separation leads to physiological deregulation, abnormal enlargement of stress-sensitive brain regions, rocking behavior, self-mutilation, anxiety, disorganized attachment styles, pathological emotional development, and, of course, petrifying nightmares. Chernus looks at the psychic disintegration that occurs in traumatized chimpanzees, especially those forced apart from their mothers at an early age. Her analysis includes the personal histories of five chimpanzees (Romie, Waty, Sara, Nico, and Panco) who were un-mothered, sold into captivity, abused, and eventually rescued by the MONA sanctuary in Northern Spain. The effects of maternal separation varied from individual to individual, probably due to factors such as "innate personality differences; the age of the chimpanzee when split from mother;

the duration, nature, and severity of abuse and neglect; and the sensitivity of caretakers in introducing them gradually to their conspecifics and their new physical environment" ([2008], p. 458).

75 Dudai (2004).

76 Quoted in Hacking (2001), pp. 252–53.

77 Aristotle described lucid dreaming in the fourth century BCE in his treatise *On Dreams*, but it was not until the nineteenth century that people began taking it seriously as a mental phenomenon, thanks primarily to the writings of two French authors: Hervey de Saint-Denys (1822–1892) and Alfred Maury (1817–1892). The term "lucid dream" was then coined by the Dutch psychiatrist Frederik van Eeden at the start of the twentieth century. He used it to refer to those rare dream experiences whose most prominent characteristic is that they do ***not*** impair us metacognitively. After meticulously logging and analyzing 500 personal dreams, Eeden discovered that a significant portion of them—352 out of 500, to be exact—were "of a particular kind." In them, he says, "I had full recollection of my day- life, and could act voluntarily, though I was fast asleep." In an essay published in the *Proceedings of the Society for Psychical Research* in 1913, Eeden argued that these dreams were different enough from typical dreams to deserve their own nomenclature. At one point, he worried that they were so different that others may doubt their existence or, worse, refuse to see them as dreams at all. "If anybody refuses to call that state of mind a dream, he may suggest some other name. For my part, it was just this form of dream, which I call 'lucid dreams,' which aroused my keenest interest and which I noted down most carefully." Eeden's worries were well-founded since, due to its "fantastic" nature, lucid dreaming was relegated to the margins of science. Only in the late 1970s and early 1980s did scientific attitudes begin to change, thanks in large part to the publication of two books that brought the phenomenon back within the fold of scientific discourse: Keith Hearne's *Lucid Dreams: An Electrophysiological and Psychological Study* (1978) and Stephen LaBerge's *Lucid Dreaming* (1985). These books convinced scientists that lucid dreaming is a real phenomenon that can be studied, controlled, and manipulated in a laboratory setting.

78 Windt and Voss (2018), p. 388.

79 Walsh and Vaughan (1992), p. 196.

80 Filevich et al. (2015), p. 1082.

81 Kahan (1994), p. 251.

82 Voss and Hobson (2014), p. 16.

83 Although Voss and Hobson's position lacks detail, it repeats an argument that many philosophers of mind have made about the relationship between language and thought, which is that language allows us to form abstract mental concepts (for example, *tree*) that exceed the concrete particulars given through the doors of the senses (such as *this* pine tree, *this* sycamore, *this* weeping willow). These concepts, in turn, give us the power to form mental judgments of the form "X is Y" by subsuming particulars under universals (such as "This, here, is a tree"). It seems that in Voss and Hobson's view, the only difference between a dreamer and a lucid dreamer is that the latter forms a particular mental judgment during a dream: "This, here, is a dream."

84 For Windt and Metzinger (2007), C-lucidity is A-lucidity plus an additional component. As such, all cases of C- lucidity logically entail A-lucidity, but the inverse is not true.

85 On animal conceptuality, consult Allen (1999), Glock ([1999], [2000], [2010]), Stephan (1999), and Newen and Bartels (2007). On animal logic, see Hurley and Nudds (2006), Watanabe and Huber (2006), Call (2006), Allen (2006), Erdőhegyi et al. (2007), Schmitt and Fischer (2009), Pepperberg (2013), and Felipe de Souza and Schmidt (2014). On animal mathematics, see Boysen and Hallberg (2000), Olthof and Roberts (2000), West and Young (2002), Kilian et al. (2003), Harris, Beran, and Washburn (2007), Aust et al. (2008), Matsuzawa (2009), Rescorla (2009), Uller and Lewis (2009), Dadda et al. (2009), Pepperberg (2012), and Vonk and Beran (2012).

86 Windt and Metzinger (2007), p. 222.

87 Imagine that I see a blurry object in the distance while taking a walk through the park. At first, I may be able to tell that it is an object, but not what kind of object it is. My experience of it is vague and imprecise. Is it a bicycle or a man? Is it a statue or a drinking fountain? As I approach it, I start attributing properties to it and, along the way, eliminating certain possibilities. The object moves, so it is neither a statue nor a drinking fountain. It has a face, so it is not a bicycle. So, I think to myself: "It's a man!" According to philosophers of mind, this thought—"it is a man"—is a judgment because it has a propositional structure with a subject ("this") and a predicate ("is a man"), which indicates that I am mentally subsuming a particular (the specific object I see) under a universal (the concept "man") through the copula ("is").

88 I understand that it may seem paradoxical to speak of a pre-cognitive form of metacognition, but the idea is simply that there are forms of conscious awareness in which the subject reflects—or monitors—their own mental states without this reflection having a linguistic form or requiring advanced conceptual content.

89 To consider how this pre-conceptual experience of lucidity could come about, recall that one of the structural properties of the waking field is *transparency*. When we are awake, our perceptual field is transparent in the sense we are not aware of it as a field. The field "auto-conceals." In lucid dreams, however, this field loses this transparency. It suddenly "stands out," asserting itself as an object of our perception. When this happens, our perceptual field becomes opaque in the sense that we stop looking through it and start looking *at* it. If animals get the feeling in the middle of a dream that they are looking at, rather than through, their perceptual field, their experience would be one of lucidity.

90 Rowlands (2009), p. 210.

91 Here, I have focused on Windt and Metzinger's theory, but other experts on lucidity have distinguished between the *felt* moment of lucidity and the subsequent *judgment* of it. LaBerge and DeGracia (2000) explain that lucidity happens in two stages. First, we undergo a moment of metacognitive insight, which is the "direct experiential realization of, and self-reflection on, [our] condition." Then, we subsume this insight into a semantic framework by interpreting our condition *as* an instance of lucid dreaming. While these stages are usually co-present in human lucid dreams, they can be dissociated. A dreamer can realize that something is "off" about their condition without plugging this realization into higher-order epistemic or hermeneutic operations. Animals' lack of language or conceptuality, therefore, need not be an obstacle to them experiencing lucidity. As Metzinger argues, even without the power to "categorize" their dream experiences as dreams, people can experience lucidity "in terms of regaining agency and a stable attentional PMIR [phenomenal model of the intentionality relation] during the dream state" ([2003], p. 532).

92 Smith and Washburn (2005); Kornell, Son, and Terrace (2007); Smith (2009); Call (2010); Smith, Couchman, and Beran (2012).

93 Thompson (2015), p. 158. Thompson focuses on the dorsolateral prefrontal cortex, but a recent study by Baird, Mota-Rolim, and Dresler suggests that lucid dreaming depends on interconnectivity among multiple brain

structures. The authors note, however, that lucid dreaming may be multiply realizable at the neural level since it may be brought about by vastly different neural circuits ([2019], p. 12). This means that we cannot assume that animals without brain structures homologous or even analogous to those of humans cannot dream lucidly.

94 Hobson and Voss (2010), p. 164. Hobson and Voss do not incorporate into their analysis Windt and Metzinger's (2007) distinction between A-lucidity and C-lucidity, but they define lucidity as "insight into the fact that one is dreaming" ([2010], p. 155), which is how Windt and Metzinger define A-lucidity. Using Edelman's ([2003], [2005]) distinction between primary and secondary consciousness, they present dreaming as an expression of primary consciousness that is marked by "simple awareness" (in other words, by perceptual and emotional experiences). By contrast, they see waking experience as a combination of primary and secondary consciousness, which is to say, as a mixture of simple awareness and "awareness of awareness" (in other words, of metacognition). At first, Hobson and Voss extend lucidity only to nonhuman primates, but they eventually concede that birds are likely candidates as well. In a separate publication, Voss mentions primates again, but not birds ([2010], p. 52).

95 Manger and Siegel (2020), p. 2.

96 Pantani, Tagini, and Raffone (2018), p. 176.

97 Foucault (1985), p. 53.

98 Foucault (1985), p. 45.

99 Foucault (1985), p. 53.

100 Cyrulnik (2013), p. 143.

101 Sartre (2004), p. 15.

102 Lucretius (1910), p. 170.

CHAPTER 3

1 Coleridge (2004), p. 123.

2 Thomas's model has three axes: "absence-presence (which might more perspicuously be called stimulus constrainedness), will (or amenability to voluntary control), and the old Humean dimension of 'vivacity' or vividness" ([2014], p. 159). Thus, dreams will be closer to the "absence" end of the first axis than percepts since they do not rely on external stimuli, and on the

lower end of the "will" axis than intentional acts of imagination since they are not under voluntary control (unless they are lucid dreams, in which case they will be on the higher end). Similarly, dreaming may be more vivid than remembering but less than perceiving.

3 Foucault (1985), p. 40. Not everyone, though, agrees that dreams are imaginative exercises. Proponents of the "orthodox view" maintain that dreams are *beliefs* formed during sleep. The most famous proponent of this view is René Descartes. This is challenged by two alternative views. The "hallucinatory view" holds that dreams are, phenomenologically speaking, much closer to hallucinations than beliefs. They are richer in imagery than beliefs, and they immerse us in a perceptual reality that beliefs do not. Alternatively, the "imagination view" maintains that dreams are even closer to imaginings than to hallucinations. In chapter 2, I cited the work of Jennifer Windt and Evan Thompson. The former adheres to the hallucination view; the latter, to the imagination view. Other adherents of the imagination view include Walton (1990), Foulkes (1999), Ichikawa (2009), and Sosa (2005). Although I prefer the imagination view, my position is that dreams, imaginings, *and* hallucinations are all part of the same spectrum of imagination.

4 In chapter 1, I pointed out that Immanuel Kant believed the faculty of creative imagination was exclusively human. This is a common trend from antiquity to the present, with few exceptions. Even philosophers who have admitted that animals have imagination usually define their imaginative capacities as impoverished compared to ours (as in the case of Aristotle, who thought that animals have only sensible imagination) or subsume them under the umbrella of instinct (as in the case of Augustine, who says that animals imagine but only instinctually). For an account of the exclusion of animals throughout the history of philosophy, see Simondon (2011).

5 Foucault (1985), p. 33.

6 Luce (1966), p. 1.

7 This protocol is cited by Shafton (1995), Hartmann (2001), Foulkes (1999), and Manger and Siegel (2020), among others.

8 Luce (1966), p. 86.

9 Even if Vaughan's hallucination study came to naught, subsequent experiments have vindicated its guiding hypothesis. Rhesus monkeys hallucinate while awake, especially under the influence of amphetamines. See Siegel (1973), Siegel, Brewster, and Jarvik (1974), Siegel and Jarvik (1975), Brower and Siegel (1977), Ellison, Nielsen, and Lyon (1981), Ridley et al. (1982),

Castner and Goldman-Rakic ([1999], [2003]), and Visanji et al. (2006). Hallucinations have also been reported in rats, pigeons, and cats (Ellinwood, Sudilovsky, and Nelson [1973]). For a review of this literature, see Robbins (2017).

10 Luce (1966), p. 86.

11 Luce (1966), p. 85.

12 Luce (1966), p. 86.

13 Lohmar (2007), p. 58.

14 Lohmar does not deny that some primates (specifically humans) add a linguistic coating to these visual experiences by putting to work the linguistic-conceptual mode, but that is a far cry from thinking animals cannot mentally represent the world in the absence of language. Animals can represent the world and reenact scenes from their lives by taking advantage of other modalities that evolutionarily predate language ([2007], p. 61). Furthermore, although he developed his theory with primates in mind, Lohmar concedes that it may apply to all "highly cerebralized" animals. To deal with other animals, of course, we may need to incorporate modes of representation (olfactory, tactile, acoustic, etc.) that are not contained in Lohmar's original list of four.

15 Kunzendorf (2016) argues that, given everything we know experimentally about primate imagination, Vaughan's findings should not surprise us.

16 Kunzendorf (2016), pp. 38–39.

17 Lillard (1994) defines pretense as involving six factors: there is a pretender (one) who finds herself in a specific reality (two) and who then projects (three) a mental representation (four) onto this reality with intention (five) and awareness (six).

18 Lyn, Greenfield, and Savage-Rumbaugh (2006), p. 208.

19 Kunzendorf (2016), p. 39.

20 Matsuzawa (2011), p. 133. Lyn, Greenfield, and Savage-Rumbaugh report another interesting behavior. When they asked the chimpanzee Panbanisha to feed grapes to a toy puppet, she put the bowl of grapes to the puppet's mouth with one hand and held it there. Then, with her other hand, she guided the puppet's head into the bowl, "as if making it eat" ([2006], p. 208). "The action of moving the puppet's head to indicate pretend 'eating' may reveal an ability to extend the carer's initial pretend stipulation to feed the puppet [. . .] Such extensions imply an understanding of the pretend nature of the game" (p. 208). Another chimpanzee in the study, Panpanzee,

"groomed" the same puppet and pretended to eat the bugs he "picked" from the puppet's body. Gómez and Martín-Andrade (2005) describe many other instances of fantasy play among apes. Evidence of pretense in primates dates to the start of the twentieth century (Kinnaman [1902]). For a review of the evidence of object play among apes, see Ramsey and McGrew (2005).

21 The subfield of *Pan*-thanatology has developed around reports like this one that suggest that primates understand death. I discuss some of this literature in Peña-Guzmán (2017).

22 Mitchell (2016), pp. 333–34.

23 Writing about the experiments on chimpanzees conducted by the Gestalt psychologist Wolfgang Köhler in the first half of the twentieth century, the philosopher Peter Carruthers argues that the chimpanzees came up with their innovative solutions to the spatial problems Köhler threw their way without a shred of conscious awareness (Carruthers [1996]). "Puzzling" is what Mitchell (2016) diplomatically calls Carruthers' argument.

24 A key figure in Mitchell's history is the German philosopher and evolutionary theorist Karl Groos, who applied evolutionary principles to animal play. In *The Play of Animals* (1898), Groos explains that playful behavior is unintentional mimicry of adult behavior that helps young animals prepare for the future. But, as Mitchell points out, not all animals who play are young. Aware of this problem, Groos claimed that playful behavior in adults must be understood as *intentional make-believe* since adults do not need to practice behaviors related to survival given that their status as adults presupposes they have already mastered them. Groos noticed something that few experts on animal pretense talk about these days, which is the "pleasurable quality" of shuttling back and forth between the real and the imaginary. In noticing this quality, Groos followed in Darwin's footprints. A few decades earlier, Darwin had written in *The Descent of Man* that animals sometimes use pretense for the simple joy of watching their victims struggle (animal *schadenfreude*). It was against Groos's theory of animal play that C. Lloyd Morgan would later articulate his famous canon, which drove scientists to abandon talk of animal mentation.

25 This dolphin behavior, which was first reported by Tayler and Saayman (1973), goes beyond rote repetition. It involves, according to Kunzendorf, "visual imagination of a transformational nature" ([2016], p. 39).

26 The philosopher Kendall Walton defines dreams as a subcategory of pretense, as "games of make-believe" (1990). In *Mental Evolution in Animals*,

Romanes also presented dreaming and pretending as comparable instances of imagination.

27 Bekoff and Jamieson (1991), p. 20.

28 This is corroborated by O'Neill, Senior, & Csicsvari (2006) and by O'Neill et al. (2008).

29 Foster and Wilson (2006), p. 680.

30 Davidson, Kloosterman, and Wilson (2009), p. 504.

31 Karlsson and Frank (2009), p. 2 (my italics).

32 Karlsson and Frank write, "The mnemonic content of awake replay can be effectively independent of location" ([2009], p. 7).

33 Gupta et al. (2010), pp. 695–96.

34 Gupta et al. (2010), p. 702.

35 Derdikman and Moser (2010), p. 584.

36 Davidson, Kloosterman, and Wilson (2009), p. 503.

37 Knierim (2009), p. 422. Davidson, Kloosterman, and Wilson (2009) make the same claim about remote replay, stating that it is not "tied to the animal's current location" (p. 502).

38 Knierim (2009), p. 422.

39 At these critical junctures, rats temporarily suspend their active engagement with the world and consciously plan their trajectory by "vicariously" reconstructing all the potential paths in front of them before committing to one. Johnson and Redish (2007) point out that the neural sweeps they observed at these "high-cost choice points" occurred on the way *from,* not *to,* the animals' location, indicating that the animals were planning future action rather than remembering past behavior. They write, "Reconstruction in front of but not behind the animal suggests that the information is related to representation of future paths rather than a replay of recent history" (p. 12183). They also note that these sweeps are accompanied by fast lateral head movements as the animals look at the different alternatives again and again while thinking about them. These sweeps give animals critical information about *possible* futures. They provide them "with a prediction of the consequences of [their] action, from which an evaluation of the goal can be reached, and a decision made" (p. 12184).

40 Knierim (2009), p. 421. Karlsson and Frank say that their findings occur on too short a timescale to match the time scale of lived experience but note that others have found replay events in the hippocampus that match the time scale of waking life ([2009], p. 7). Beyond rats, Willett (2014)

describes a similar phenomenon observed by the primatologist Barbara Smuts. Smuts saw a troop of baboons in Gombe National Park in Tanzania become completely silent upon reaching a pond. All the animals, even the noisy juveniles, fell into "silent contemplation." Smuts interpreted this as a form of "baboon *shanga*," a Sanskrit word meaning communal association. Building on this interpretation, Willet speaks of it as "enlightened repose" (p. 102).

41 Lohmar's 2016 book has not yet been translated into English. The full German title is *Denken ohne Sprache. Phänomenologie des nicht-sprachlichen Denkens bei Mensch und Tier im Licht der Evolutionsforschung, Primatologie und Neurologie.*

42 Examples include Montangero (2012), Occhionero and Cicogna (2016), Domhoff (2017), and Eeles et al. (2020).

43 The few scientists and philosophers who write about animal imagination tend to see it as mammalian. But if dreams are imaginative acts, the dreams of nonmammals would challenge this view. Creativity and imagination appear to be qualities that mammals share at least with birds (Ackerman [2016]).

44 Hills (2019), p. 1.

CHAPTER 4

1 Siewert (1994), p. 200.

2 Chalmers (2018), p. 12.

3 Griffin (1976), p. 15.

4 Luce (1966), p. 1.

5 Rowlands (2009), p. 176.

6 Bekoff and Jamieson (1991), p. 15.

7 There is a strand of environmental ethics that presses against this view by extending moral and legal standing to nonliving entities, such as trees and rivers. In the case of nonliving beings, the argument turns on legal rather than moral status. See Brennan (1984) and Stone (2010).

8 Warren (1997), p. 3 (also quoted in Shepherd [2018], p. 14).

9 Jewish philosophers have written about ethics in ways that resonate with this way of thinking. One example is the dialogical ethics of Martin Buber (see Buber [1970]).

10 Block (1995), p. 231.

11 Block (1995), p. 230.
12 Levy (2014), p. 128.
13 Searle (1997), p. 98. This is not to say that pain is entirely without cognition, as if all our experiences of pain were isolated sensations that remain unaffected by our emotional states, our previous beliefs, or even our social and cultural contexts. The intensity of stubbing my toe is not the same when I am sleepy and the pain wakes me up in a jolt, and when I am so engrossed in a task that I barely flinch. Social, cognitive, psychological, and existential factors mold the meaning of pain. That being said, my point is that access consciousness cannot deplete the phenomenology of pain because there is a dimension to my experiences of pain that no cognitive account can capture. This is the qualitative dimension of how my pain *feels* to me, of how I experience it in a particular moment. Block uses pain to illustrate that our lived experience of the world carries a reminder that cannot be captured with the net of access consciousness. For a critique of representationalist theories of pain, see the work of Colin Klein.
14 This argument appears in Siewert (1998), which builds upon Siewert (1994).
15 Siewert (1994), p. 216. There are some philosophers who believe that Block's notion of phenomenal consciousness is incoherent, but research in cognitive psychology and experimental philosophy indicates that something very similar to Block's distinction between access and phenomenal consciousness is baked into commonsense psychology (Knobe and Prinz [2008]). According to Huebner, "There are distinctions, perhaps implicit in the structure of commonsense judgments, that seem to map something like a distinction between phenomenal and nonphenomenal mental states" ([2010], p. 135).
16 Joshua Shepherd defends a similar thesis in his book *Consciousness and Moral Status*. Phenomenal consciousness has intrinsic value because it gives birth, in all species who have it, to an "evaluative space" that allows organisms to evaluate things as either positive or negative, as attractive or repulsive ([2018], p. 73). By giving rise to this space, phenomenal consciousness makes possible the kinds of affective and evaluative experiences that make life worthwhile. When deciding whether an entity has moral value, therefore, ethicists must pay attention not to "the entity's *smarts*, or whether it is rational, or self-aware" (p. 92). They need to pay attention to its *feels*, to whether it evaluates things and has a point of view. In the second to last chapter, entitled "Moral Status: The Other Animals," Shepherd applies

his theory of phenomenal value to animals, concluding that moral value probably applies to many more animals than we expect. Those who cannot escape "the philosopher's yen for line-drawing," he says, "[have no choice but] to draw the line quite low on the evolutionary totem pole" (p. 99).

17 Organisms do not have to be conscious of the fact that they perform acts of valuing in order for these acts to fill their existence with meaning. As soon as organisms express a preference—say, by seeking attractive stimuli and moving away from repellant ones—they have performed an evaluative act since they have attached a valence to a stimulus.

18 The most famous application of consequentialist principles to animal ethics is Singer (1995).

19 One could argue that since the alpha and omega of the consequentialist's moral universe (namely pain and pleasure) are themselves phenomenal states, the basis of moral status is phenomenal consciousness. Adherents to *hedonic consequentialism* embrace this position (or some version of it), while defenders of *preference consequentialism* reject it. Preference consequentialists give less moral weight to our experience of pain and pleasure and more to the satisfaction of our preferences. For many of them, the access-first approach is more appealing since, in their view, only beings who are access conscious can form preferences. Here I focus on this brand of consequentialism since it is the one that poses the greater challenge to my position.

20 Rallying round this access-first approach, Kahane and Savulescu (2009) and Levy and Savulescu (2009) argue that subjective preferences based on "sapience" trump those based on "sentience." Levy (2014) follows suit. He deflates Siewert's warnings about the dangers of zombification by saying that the loss of phenomenal consciousness would be far less tragic than Siewert believes. According to Levy, Siewert may be right that there are many aspects of our experience that we ought to treasure (such as our experience of colors, aesthetic and sensual pleasure, interpersonal intimacy, a sense of self, and even pain), but he is wrong in imputing these to phenomenal consciousness. In reality, they depend on access consciousness. My zombie-self, Levy says, would still perform all the functions of access consciousness (thinking, acting, speaking, etc.) and all the experiences that phenomenality-first theorists mistakenly attribute to phenomenal consciousness. So zombie life, it seems, is less tragic and depleted than others think. Unfortunately, Levy deflates Siewert's position only by arbitrarily

transposing onto access consciousness everything that Siewert—and, really, most experts who work in this area—attributes to phenomenal consciousness. His assertion that my zombie-self would experience pain, enjoy the beauty of art, appreciate the joys of friendship, and experience sexual pleasure *in spite of lacking phenomenal consciousness* is incoherent since, by definition, zombies have no inner life (no phenomenology). What kind of zombie has a sense of self? What kind of zombie cultivates friendships and goes to the MOMA to admire an O'Keeffe or a Giacometti? What kind of zombie orgasms? One cannot but worry that Levy has misunderstood the core distinction between phenomenal and access consciousness and, perforce, failed to recognize the noncognitive dimensions of the phenomenal states he discusses.

21 Levy and Savulescu (2009), p. 367.

22 Levy and Savulescu (2009), p. 367.

23 Kahane and Savulescu (2009), p. 21 (italics in the original).

24 In *Cognitive Disability and Its Challenge to Moral Philosophy*, disability philosophers Eva Feder Kittay and Licia Carlson point out that the consequentialist ranking of pleasures is rooted in a deadly mixture of ignorance and prejudice (see Kittay and Carlson [2010] as well as Kittay [2009]). The same holds for the access-first approach to moral status. Proponents of this approach have defended moral positions that trouble me deeply. Julian Savulescu has made a career in bioethics out of arguing that we have a moral obligation to actively prevent the birth of people with cognitive disabilities and low IQ. I am not the first to detect clear echoes of eugenic ideology in his writings. For a criticism of Savulescu's ableism, see Sparrow (2010) and Hall ([2016], pp. 20–23).

25 Aside from having a cracked moral compass, modern disciples of Mill face a pernicious epistemological problem. In boasting that he would rather be a human dissatisfied than a pig satisfied, what can Mill possibly know about what it is like to be a happy pig? Not much, I assume. Likewise, in claiming that the lives of animals and people with disabilities offer objectively fewer goods (or less *good* goods) than the lives of neurotypical humans, on what epistemic ground do Levy, Kahane, and Savulescu stand? What authorizes them to speak with such confidence about forms of life they do not embody, life-worlds they do not inhabit? The problem cannot be sidestepped. Who is to say that being a Socrates dissatisfied truly is better than being a fool satisfied? Mill, the philosopher, has an answer. Maybe the fool has another.

26 Quoted in Shepherd (2018), pp. 98–99. Shepherd could have just as easily cited Hellenistic philosopher Epicurus, who preached that the key to happiness (*eudaimonia*) does not depend on the exercise of theoretical reason (*theoria*) as Aristotle preached, but in the attainment of simple tranquility (*ataraxia*)—in a life lived in the absence of unnecessary disturbance.

27 For Kantian approaches to animal ethics, see O'Neill (1997) and Korsgaard (2018).

28 Kriegel (2017), p. 127.

29 Kriegel (2017), p. 127.

30 Kriegel borrows the image of the weather watchers from Strawson (2012).

31 Kriegel (2017), p. 127.

32 Kriegel (2019), p. 516.

33 Many Kantians adopt this position, distinguishing between *moral agents* and *moral patients*.

34 Kriegel (n.d.), p. 27.

35 Kriegel's analysis has a distinctly Levinasian flavor to it. In books such as *Totality and Infinity: An Essay on Exteriority* and *Otherwise than Being, or Beyond Essence*, the Jewish philosopher Emmanuel Levinas develops an ethical philosophy anchored in the radical otherness—or, to use his preferred term, *alterity*—of the Other. This otherness renders the Other elusive. The Other is "an infinity" that I cannot subsume under any of my categories and to whom I am nonetheless morally accountable. See Levinas (1979) and (1981).

36 Kriegel (2017), p. 31.

37 Kriegel (2017), p. 133.

38 Kriegel (2017), p. 131.

39 In *The Republic,* Socrates tells Glaucon that lawless pleasures "are to be found in us all." "What desires do you mean?" Glaucon asks, to which Socrates responds: "Those that are awakened in sleep when the rest of the soul, the rational, gentle and dominant part, slumbers, but the beastly and savage part, replete with food and wine, gambols and, repelling sleep, endeavors to sally forth and satisfy its own instincts. You are aware that in such cases there is nothing it will not venture to undertake as being released from all sense of shame and all reason. It does not shrink from attempting to lie with a mother in fancy or with anyone else, man, god or brute. It is ready for any foul deed of blood; it abstains from no food, and, in a word, falls short of no extreme of folly and shamelessness" (Plato [2000], 571d).

40 I owe this reference to Driver (2007).

41 The issue of the connection between dream content and morality reappeared in the 1980s when the prominent journal *Philosophy* published two articles, Matthews (1981) and Mann (1983), about whether people are morally responsible for their dreams. For a discussion of these articles, see Driver (2007).

42 Sperling tested people's working memory by showing them cards with three rows of letters and asking them to recall the top, middle, or bottom row. Subjects could successfully recall (that is, access) any one of the rows, but not *all* the rows. As soon as they recalled a specific row, their cognitive access to the others vanished. Block interpreted this to mean that in the interval between stimulus and recall the subjects visually grasped the card as a whole (as an "icon"), but they did not yet have access to its component parts. Subjects were phenomenally conscious but not access conscious of the rows. Block writes: "Here is the description I think is right and that I need for my case: I am P-conscious of all (or almost all—I will omit this qualification) the letters at once, that is, jointly, and not just as blurry or vague letters, but as specific letters (or at least specific shapes), but I don't have access to all of them jointly, all at once" ([1995], p. 244).

43 Sebastián (2014a), p. 278.

44 Sebastián (2014a), p. 276.

45 The dreaming brain is busy immersing us in an imagined world analogue. Yet none of this activity unfolds in the dlPFC. Most of it occurs in the areas where basic sensory percepts are formed, such as the primary visual and auditory cortices.

46 Sebastián explains:

Now, if the dlPFC plays a fundamental role in cognitive access, as I have been arguing, an increase in its activity during lucid dreams is to be expected and would further support my claim. Preliminary empirical evidence for this hypothesis has been obtained via several studies. For example, Wehrle et al. (2005) and Wehrle et al. (2007), where fMRI was used to study brain region activation during lucid dreams, show that in lucid dreams not only frontal but also temporal and occipital regions are highly activated in comparison to non-lucid dreams. Voss et al. (2009) shows that lucid dreaming by trained participants is associated with increased electroencephalography (EEG) power, especially in the 40-Hz range, over frontal regions. Finally, Dresler et al. (2012) have published neural correlates of lucid dreams obtained from contrasting lucid and non-lucid REM sleep. Not surprisingly, the dlPFC

(Brodmann's area 46) is among the areas in which a significant increase in activity is recorded. ([2014a], p. 278).

47 In defending a phenomenalist interpretation of dreams, Sebastián ([2014a], pp. 276–77) positions himself alongside a number of famous Western philosophers who have done the same, including Immanuel Kant, Bertrand Russell, G.E. Moore, and Sigmund Freud (see also Sebastián [2014b]). He also joins a flotilla of contemporary dream researchers who share this interpretation, such as Ichikawa (2009), Ichikawa and Sosa (2009), Metzinger ([2003], [2009]), Revonsuo (2006), and Sosa (2005). Pantani et al. (2018) also share this interpretation of dream phenomenology, except that they believe dreams are created in what Damasio (1989) calls *convergence-divergence zones* in the brain, especially those responsible for the integration of sensory experiences before they become available for central processing; in other words, before they pass the informational bottleneck leading to Baars's (1997) "global workspace."

48 Examples include Bernard Baars's global workspace theory and Stanislas Dehaene's theory of conscious access.

49 Examples include David Rosenthal's theory of metacognition and Michael Tye's "PANIC" theory of consciousness.

50 A central question that appears constantly in this literature is whether moral status is an all-or-nothing affair or whether it is a matter of degree, with different animals having more or less of it. If it is an all-or-nothing business, who are the "haves" and the "have-nots"? And if it comes in degrees, how do we decide the degree of moral status that different species, or even different members of the same species, merit? How is this status measured and allocated? DeGrazia (2009) distinguishes between two ways of asserting that moral status comes in degrees. One is the *two-tier model*, according to which all humans have full moral status, and all the other animals have a smaller portion. Alternatively, there is the *sliding-scale model*, which admits that different animals have different degrees of moral status depending "on the sorts of beings they are."

51 DeGrazia (1991), p. 49.

52 Gruen (2017), p. 1.

53 I am here thinking of people such as Peter Carruthers, R.G. Frey, and Joseph LeDoux. Carruthers, for instance, moves very quickly from his claim that animals are not conscious to his claim that they make "no rational claim upon our sympathy" ([1989], p. 268).

54 Looking back at the history of the modern animal rights movement, Marc Bekoff and Dale Jamieson explain that the reason Peter Singer's *Animal Liberation* proved so influential in the closing decades of the twentieth century is because post-behaviorist developments in psychology and cognitive science had already cleared the way for its reception. It is not that these developments confirmed Singer's position. It is that they shifted our collective perspective about the minds of animals, rendering Singer's position culturally intelligible. Had those developments not taken place, Singer's message perhaps would have fallen flat, having no cultural ground to stand on.

EPILOGUE

1 Hacking (2004), p. 233.
2 Cavalieri (2003).
3 In *On the Interpretation of Dreams*, Freud distinguishes between three kinds of dreams based on content. There are phenomenologically coherent and existentially normal dreams, which depict mundane situations that we could experience in real life and that we would have no trouble folding into the story of our life upon waking (such as when I dream of teaching a class). There are phenomenologically coherent but existentially strange dreams, whose phenomenal content may be "sensibly linked together" but whose narrative content does not cohere with our self-understanding (such as when I dream of having sex with a family member). The scene is perfectly ordered, but I struggle to identify the desires expressed in the dream as mine. I do not see that dream as fitting into my self- understanding. Finally, there are dreams that are incoherent all around (for example, when I dream that I am a flying, ten-legged lizard that is also the King of Russia and then suddenly I am also a bear). Freud was particularly interested in the last two kinds of dreams.
4 Guardia (1892), p. 226.
5 Hartmann (2008), p. 53.
6 Wolfe (2013), p. 94.
7 Hobson explains that dream states embody the "auto-creative character" that animates all subjective experience in its simplest and purest form, making them "functionally superior" to waking states ([2001], p. 9).

8 In an essay entitled "Animal Life and Phenomenology," the Spanish philosophers Javier San Martín and Maria Luz Pintos Peñarada argue that although the phenomenological philosophical tradition has historically privileged the study of human experience, we can extend the phenomenological concept of subjectivity to animals because the latter are "constituting beings" (San Martín and Pintos Peñaranda [2001]).

9 Pearson and Large (2006), p. 119.

10 Pearson and Large (2006), p. 115.

11 Quoted in Domash (2020), p. 108.

REFERENCES

Aaltola, E. (2010). Animal minds, skepticism, and the affective stance. *Teorema: Revista Internacional de Filosofía* 20: 69–82.

Ackerman, J. (2016). *The Genius of Birds*. New York: Penguin.

Adrien, J. (1984). Ontogenese du sommeil chez le mammifere. In *Physiologie du sommeil*, Benoit, O. (ed.), 19–29. Paris: Masson.

Allen, C. (1999). Animal concepts revisited: the use of self-monitoring as an empirical approach. *Erkenntnis* 51: 537–44.

———. (2006). Transitive inference in animals: reasoning or conditioned associations. In *Rational Animals?*, Hurley, S. and Nudds, M. (eds.), 175–85. Oxford: Oxford University Press.

Andrews, K. (2014). *The Animal Mind: An Introduction to the Philosophy of Animal Cognition*. New York: Routledge.

Aust, U., Range, F., Steurer, M. and Huber, L. (2008). Inferential reasoning by exclusion in pigeons, dogs, and humans. *Animal Cognition* 11: 587–97.

Austin, J. H. (1999). *Zen and the Brain: Toward an Understanding of Meditation and Consciousness*. Cambridge: MIT Press.

Baars, B. J. (1986). *The Cognitive Revolution in Psychology*. New York: Guilford Press.

———. (1997). In the theatre of consciousness: global workspace theory, a rigorous scientific theory of consciousness. *Journal of Consciousness Studies* 4: 292–309.

Bachelard, G. (1963). *Le matérialisme rationnel*. Paris: Presses Universitaires de France.

Baird, B., Mota-Rolim, S. A., and Dresler, M. (2019). The cognitive neuroscience of lucid dreaming. *Neuroscience & Biobehavioral Reviews* 100: 305–23.

Balcombe, J. (2010). *Second Nature: The Inner Lives of Animals*. New York: Macmillan.

Bekoff, M. (2003). Consciousness and self in animals: some reflections. *Zygon* 38: 229–45.

Bekoff, M., and Jamieson, D. (1991). Reflective ethology, applied philosophy, and the moral status of animals. In *Perspectives in Ethology: Human Understanding and Animal Awareness*, Bateson, P. G., and Klopfer, P. H. (eds.) 1–32. New York: Plenum Press.

Bender, K. (2016). What is your cat or dog dreaming about? A Harvard expert has some answers. *People Magazine.* October 13, 2016. https://people.com/pets/what-is-your-cat- or-dog-dreaming-about-a-harvard-expert-has-some-answers/.

Bendor, D., and Wilson, M. A. (2012). Biasing the content of hippocampal replay during sleep. *Nature Neuroscience* 15: 1439–44.

Bentham, J. (1843). *The Works of Jeremy Bentham*, Bowring, J. (ed.). London: William Tait.

Berardi, A., Trezza, V., Palmery, M., Trabace, L., Cuomo, V., and Campolongo, P. (2014). An updated animal model capturing both the cognitive and emotional features of post-traumatic stress disorder (PTSD). *Frontiers in Behavioral Neuroscience* 8: 1–12.

Berger, R. J., and Walker, J. M. (1972). Sleep in the burrowing owl (*Speotyto cunicularia hypugaea*). *Behavioral Biology* 7: 183–94.

Berntsen, D., and Jacobsen, A. S. (2008). Involuntary (spontaneous) mental time travel into the past and future. *Consciousness and Cognition* 17: 1093–1104.

Block, N. (1995). On a confusion about a function of consciousness. *Behavioral and Brain Sciences* 18: 227–47.

Blumberg, M. S. (2010). Beyond dreams: do sleep-related movements contribute to brain development? *Frontiers in Neurology* 1: 140.

Bogzaran, F., and Deslauriers, D. (2012). *Integral Dreaming: A Holistic Approach to Dreams*. Albany: SUNY Press.

Botero, M. (2020). Primate orphans. In *Encyclopedia of Animal Cognition and Behavior*, Vonk, J., and Shackelford, T. K. (eds.), 1–7. New York: Springer International Publishing.

Boyce, R., Glasgow, S. D., Williams, S., and Adamantidis, A. (2016). Causal evidence for the role of REM sleep theta rhythm in contextual memory consolidation. *Science* 352: 812–16.

Boysen, S. T., and Hallberg, K. I. (2000). Primate numerical competence: Contributions toward understanding nonhuman cognition. *Cognitive Science* 24: 423–43.

Bradshaw, G. A. (2009). *Elephants on the Edge*. New Haven: Yale University Press.

Brennan, A. (1984). The moral standing of natural objects. *Environmental Ethics* 6: 35–56.
Brereton, D. P. (2000). Dreaming, adaptation, and consciousness: the social mapping hypothesis. *Ethos* 28: 379–409.
Brower, K. J., and Siegel, R. K. (1977). Hallucinogen-induced behaviors of free-moving chimpanzees. *Bulletin of the Psychonomic Society* 9: 287–90.
Buber, M. (1970). *I and Thou*. New York: Scribner.
Burgin, C. J., Colella, J. P., Kahn, P. L., and Upham, N. S. (2018). How many species of mammals are there? *Journal of Mammalogy* 99: 1–14.
Calkins, M. W. (1893). Statistics of dreams. *American Journal of Psychology* 5.3: 311–43.
Call, J. (2006). Inferences by exclusion in the great apes: The effect of age and species. *Animal Cognition* 9: 393–403.
———. (2010). Do apes know that they could be wrong? *Animal Cognition* 13: 689–700.
Campbell, R. L., and Germain, A. (2016). Nightmares and posttraumatic stress disorder (PTSD). *Current Sleep Medicine Reports* 2: 74–80.
Carruthers, P. (1989). Brute experience. *Journal of Philosophy* 86: 258–69.
———. (1996). *Language, Thought and Consciousness: An Essay in Philosophical Psychology*. Cambridge: Cambridge University Press.
———. (2008). Meta-cognition in animals: A skeptical look. *Mind & Language* 23: 58–89.
Carson, A. (1994). The glass essay. *RARITAN* 13: 25.
Castner, S. A., and Goldman-Rakic, P. S. (1999). Long-lasting psychotomimetic consequences of repeated low-dose amphetamine exposure in rhesus monkeys. *Neuropsychopharmacology* 20.1: 10–28.
Castner, S. A., and Goldman-Rakic, P. S. (2003). Amphetamine sensitization of hallucinatory-like behaviors is dependent on prefrontal cortex in nonhuman primates. *Biological Psychiatry* 54: 105–10.
Cavalieri, P. (2003). *The Animal Question: Why Nonhuman Animals Deserve Human Rights*. Oxford: Oxford University Press.
———. (2012). Declaring whales' rights. *Tamkang Review* 42: 111–37.
———. (2018). The meta-problem of consciousness. *Journal of Consciousness Studies* 25: 6–61.
Chase, M. H., and Morales, F. R. (1990). The atonia and myoclonia of active (REM) sleep. *Annual Review of Psychology* 41: 557–84.

Chernus, L. A. (2008). Separation/abandonment/isolation trauma: An application of psychoanalytic developmental theory to understanding its impact on both chimpanzee and human children. *Journal of Emotional Abuse* 8: 447–68.

Churchland, P. M. (1995). *The Engine of Reason, the Seat of the Soul: A Philosophical Journey into the Brain*. Cambridge: MIT Press.

Coleridge, S. (2004). *The Complete Poems of Samuel Taylor Coleridge*. London: Penguin.

Conn, Jacob H. (1974). The decline of psychoanalysis: The end of an era, or here we go again. *JAMA* 228.6: 711–12.

Corner, M. A. (2013). Call it sleep—what animals without backbones can tell us about the phylogeny of intrinsically generated neuromotor rhythms during early development. *Neuroscience Bulletin* 29: 373–80.

Corner, M., and van der Togt, C. (2012). No phylogeny without ontogeny—a comparative and developmental search for the sources of sleep-like neural and behavioral rhythms. *Neuroscience Bulletin* 28: 25–38.

Cortés Z. C. (2015). Nonhuman animal testimonies: a natural history in the first person? In *The historical animal*, Nance, S., Colby, J., Gibson, A. H., Swart, S., Tortorici, Z., and Cox, L. (eds.), 118–32. Syracuse: Syracuse University Press.

Crick, F., and Mitchison, G. (1983). The function of dream sleep. *Nature* 304: 111–14.

Crist, E. (2010). *Images of Animals*. Philadelphia: Temple University Press.

Cyrulnik, Boris. (2013). Les animaux rêvent-ils? Quand le rêve devient liberté. *Le Coq-Héron* 4.215: 142–49.

Dadda, M., Piffer, L., Agrillo, C., and Bisazza, A. (2009). Spontaneous number representation in mosquitofish. *Cognition* 112: 343–48.

Damasio, A. R. (1989). Time-locked multiregional retroactivation: A systems-level proposal for the neural substrates of recall and recognition. *Cognition* 33: 25–62.

———. (1999). *The Feeling of What Happens: Body and Emotion in the Making of Consciousness*. New York: Houghton Mifflin Harcourt.

Darwin, C. (1891). *The Descent of Man and Selection in Relation to Sex*. London: John Murray.

Dave, A. S., and Margoliash, D. (2000). Song replay during sleep and computational rules for sensorimotor vocal learning. *Science* 290: 812–16.

Davidson, T. J., Kloosterman, F., and Wilson, M. A. (2009). Hippocampal replay of extended experience. *Neuron* 63: 497–507.

Dawkins, M. S. (2012). *Why Animals Matter: Animal Consciousness, Animal Welfare, and Human Well-being*. Oxford: Oxford University Press.

de Waal, F. (2016). *Are We Smart Enough to Know How Smart Animals Are?* New York: WW Norton & Company.

DeGrazia, D. (1991). The moral status of animals and their use in research: A philosophical review. *Kennedy Institute of Ethics Journal* 1: 48–70.

———. (2009). Self-awareness in animals. In *The Philosophy of Animal Minds*, Lurz, R. W. (ed.), 201–17. Cambridge: Cambridge University Press.

Dehaene, S. (2014). *Le code de la conscience*. Paris: Odile Jacob.

Derdikman, D., and Moser, M. (2010). A dual role for hippocampal replay. *Neuron* 65: 582–84.

Derégnaucourt, S., and Gahr, M. (2013). Horizontal transmission of the father's song in the zebra finch (*Taeniopygia guttata*). *Biology Letters* 9: 20130247.

Derrida, J. (2002). The animal that therefore I am (more to follow). *Critical Inquiry* 28: 369–418.

Despret, V. (2016). *What Would Animals Say If We Asked the Right Questions?* Minneapolis: University of Minnesota Press.

Dewasmes, G., Cohen-Adad, F., Koubi, H., and Le Maho, Y. (1985). Polygraphic and behavioral study of sleep in geese: Existence of nuchal atonia during paradoxical sleep. *Physiology & Behavior* 35: 67–73.

Domash, L. (2020). *Imagination, Creativity and Spirituality in Psychotherapy: Welcome to Wonderland*. New York: Routledge.

Domhoff, G. W. (2017). *The Emergence of Dreaming: Mind-wandering, Embodied Simulation, and the Default Network*. Oxford: Oxford University Press.

Driver, J. (2007). Dream immorality. *Philosophy* 82: 5–22.

Dudai, Y. (2004). The neurobiology of consolidations, or, how stable is the engram? *Annual Review of Psychology* 55: 51–86.

Dumpert, J. (2019). *Liminal Dreaming: Exploring Consciousness at the Edges of Sleep*. Berkeley: North Atlantic Books.

Duntley, S. P. (2003). Sleep in the cuttlefish sepia officinalis. *Sleep* 26: A118.

———. (2004). Sleep in the cuttlefish. *Annals of Neurology* 56: S68.

Duntley, S. P., Uhles, M., and Feren, S. (2002). Sleep in the cuttlefish sepia pharaonic. *Sleep* 25: A159–A160.

Edelman, Gerald M. (2003). Naturalizing consciousness: A theoretical framework. *Proceedings of the National Academy of Sciences* 100.9: 5520–24.

———. (2005). *Wider than the Sky: A Revolutionary View of Consciousness*. London: Penguin.

Edgar, D. M., Dement, W. C., and Fuller, C. A. (1993). Effect of SCN lesions on sleep in squirrel monkeys: Evidence for opponent processes in sleep-wake regulation. *Journal of Neuroscience* 13: 1065–79.

Eeles, E., Pinsker, D., Burianova, H., and Ray, J. (2020). Dreams and the day-dream retrieval hypothesis. *Dreaming* 30: 68–78.

Ellinwood, E., Sudilovsky, A., and Nelson, L. M. (1973). Evolving behavior in the clinical and experimental amphetamine (model) of psychosis. *American Journal of Psychiatry* 130: 1088–93.

Ellison, G., Nielsen, E. B., and Lyon, M. (1981). Animal model of psychosis: Hallucinatory behaviors in monkeys during the late stage of continuous amphetamine intoxication. *Journal of Psychiatric Research* 16: 13–22.

Erdőhegyi, Á., Topál, J., Virányi, Z., and Miklósi, Á. (2007). Dog-logic: Inferential reasoning in a two-way choice task and its restricted use. *Animal Behaviour* 74: 725–37.

Felipe de Souza, M., and Schmidt, A. (2014). Responding by exclusion in Wistar rats in a simultaneous visual discrimination task. *Journal of the Experimental Analysis of Behavior* 102: 346–52.

Filevich, E., Dresler, M., Brick, T. R., and Kühn, S. (2015). Metacognitive mechanisms underlying lucid dreaming. *Journal of Neuroscience* 35.3: 1082–88.

Fisher, N. (2017). Kant on animal minds. *Ergo, an Open Access Journal of Philosophy* 4: 441–62.

Foster, D. J., and Wilson, M. A. (2006). Reverse replay of behavioural sequences in hippocampal place cells during the awake state. *Nature* 440: 680–83.

Foucault, M. (1985). Dream, imagination, and existence. *Review of Existential Psychology and Psychiatry* 19:1: 29–78.

Foulkes, D. (1990). Dreaming and consciousness. *European Journal of Cognitive Psychology* 2: 39–55.

———. (1999). *Children's Dreaming and the Development of Consciousness*. Cambridge: Harvard University Press.

Frank, M. G. (1999). Phylogeny and evolution of rapid eye movement (REM) sleep. In *Rapid Eye Movement Sleep*, Mallick, B. N., and Inoué, S. (eds.), 17–38. New York: Narosa.

Frank, M. G., Waldrop, R. H., Dumoulin, M., Aton, S., and Boal, J. G. (2012). A preliminary analysis of sleep-like states in the cuttlefish Sepia officinalis. *PLoS One* 7: e38125.

Freiberg, A. S. (2020). Why we sleep: A hypothesis for an ultimate or evolutionary origin for sleep and other physiological rhythms. *Journal of Circadian Rhythms* 18: 2–6.

Freud, S. (1938). The interpretation of dreams. In *The Basic Writings of Sigmund Freud*, Brill, A. A. (ed.), 181–549. New York: Random House.

Gallup, G. G. (1977). Self-recognition in primates: A comparative approach to the bidirectional properties of consciousness. *American Psychologist* 32: 329–38.

Gardner, H. (1987). *The Mind's New Science: A History of the Cognitive Revolution*. New York: Basic Books.

Gardner, R. A., Gardner, B. T., and Van Cantfort, T. E., eds. (1989). *Teaching Sign Language to Chimpanzees*. Albany: SUNY Press.

Gelbard-Sagiv, H., Mukamel, R., Harel, M., Malach, R., and Fried, I. (2008). Internally generated reactivation of single neurons in human hippocampus during free recall. *Science* 322: 96–101.

Gioanni, H. (1988). Stabilizing gaze reflexes in the pigeon (*Columba livia*). *Experimental Brain Research* 69: 567–82.

Glock, H. J. (1999). Animal minds: Conceptual problems. *Evolution and Cognition* 5: 174–88.

———. (2000). Animals, thoughts and concepts. *Synthese* 123: 35–64.

———. (2010) Can animals judge? *Dialectica* 64: 11–33.

Godfrey-Smith, P. (2016). *Other Minds: The Octopus, the Sea, and the Deep Origins of Consciousness*. New York: Farrar, Straus and Giroux.

———. (2017). The mind of an octopus. *Scientific American*, January 1, 2017. https://www.scientificamerican.com/article/the-mind-of-an-octopus/.

Gómez, J. C., and Martín-Andrade, B. (2005). Fantasy play in apes. In *The Nature of Play: Great Apes and Humans*, Pellegrini, A. D., and Smith, P. K. (eds.), 139–72. New York: Guilford Press.

Graf, R., Heller, H. C., and Rautenberg, W. (1981). Thermoregulatory effector mechanism activity during sleep in pigeons. In *Contributions to Thermal Physiology*, Szelenyi, Z., and Szekely, M. (eds.), 225–27. Oxford: Oxford Press.

Graf, R., Heller, H. G., and Sakaguchi, S. (1983). Slight warming of the spinal cord and the hypothalamus in the pigeon: effects on thermoregulation and sleep during the night. *Journal of Thermal Biology*, 8.1–2: 159–61.

Griffin, D. R. (1976). *The Question of Animal Awareness: Evolutionary Continuity of Mental Experience*. New York: Rockefeller University Press.

———. (1998). From cognition to consciousness. *Animal Cognition* 1: 3–16.
Groos, Karl. (1898). *The Play of Animals*. Boston: D. Appleton and Company.
Gruen, L. (2017). The moral status of animals. *Stanford Encyclopedia of Philosophy*. Accessed July 23, 2020. https://plato.stanford.edu/entries/moral-animal/.
Guardia, J. M. (1892). La personalité dans les rêves. *Revue Philosophique de la France et de l'Étranger* 34: 225–58.
Gupta, A. S., van der Meer, M. A., Touretzky, D. S., and Redish, A. D. (2010). Hippocampal replay is not a simple function of experience. *Neuron* 65: 695–705.
Hacking, I. (2001). Dreams in place. *Journal of Aesthetics and Art Criticism* 59: 245–60.
———. (2004). *Historical Ontology*. Cambridge: Harvard University Press.
Hale, N. G., Jr. (1995). *The Rise and Crisis of Psychoanalysis in the United States: Freud and the Americans, 1917–1985*. Oxford: Oxford University Press.
Hall, M. (2016). *The Bioethics of Enhancement: Transhumanism, Disability, and Biopolitics*. Lanham, Maryland: Lexington Books.
Halton, E. (1989). An unlikely meeting of the Vienna school and the New York school. *New Observations* 1: 5–9.
Harris, E. H., Beran, M. J., and Washburn, D. A. (2007). Ordinal-list integration for symbolic, arbitrary, and analog stimuli by rhesus macaques (*Macaca mulatta*). *Journal of General Psychology* 134: 183–97.
Hartmann, E. (1995). Making connections in a safe place: Is dreaming psychotherapy? *Dreaming* 5: 213.
———. (2001). *Dreams and Nightmares: The Origin and Meaning of Dreams*. Cambridge: Perseus Publishing.
———. (2008). The central image makes "big" dreams big: The central image as the emotional heart of the dream. *Dreaming* 18: 44–57.
Haselswerdt, Ella. (2019). The Semiotics of the Soul in Ancient Medical Dream Interpretation: Perception and the Poetics of Dream Production in Hippocrates' "On Regimen." *Ramus* 48.1: 1–21.
Hearne, K.M.T. (1978). *Lucid Dreams: An Electro-physiological and Psychological Study*. Doctoral dissertation, Liverpool University.
Hernandez-Lallement, J., Attah, A. T., Soyman, E., Pinhal, C. M., Gazzola, V., and Keysers, C. (2020). Harm to others acts as a negative reinforcer in rats. *Current Biology* 30: 949–61.
Hills, T. (2019). Can animals imagine things that have never happened? *Psychology Today*. Accessed October 22, 2019. https://www.psychologytoday.com

/us/blog/statistical-life/201910/can-animals-imagine-things-have-never-happened.
Hobson, J. A. (2001). *The Dream Drugstore: Chemically Altered States of Consciousness*. Cambridge: MIT Press.
Hobson, J. A., and McCarley, R. W. (1977). The brain as a dream state generator: An activation-synthesis hypothesis of the dream process. *American Journal of Psychiatry* 134: 1335–48.
Hobson, A., and Voss, U. (2010). Lucid Dreaming and the Bimodality of Consciousness. In *New Horizons in the Neuroscience of Consciousness*, Perry, E. K., Collerton, D., LeBeau, F.E.N., and Ashton, H. (eds.), 155–68. Amsterdam: John Benjamins Publishing Company.
Huebner, B. (2010). Commonsense concepts of phenomenal consciousness: Does anyone care about functional zombies? *Phenomenology and the Cognitive Sciences* 9: 133–55.
Hurley, S. E., and Nudds, M. (2006). *Rational Animals?* Oxford: Oxford University Press.
Ichikawa, J. (2009). Dreaming and imagination. *Mind & Language* 24: 103–21.
Ichikawa, J, and Sosa, E. (2009). Dreaming, philosophical issues. In *The Oxford Companion to Consciousness*, Bayne, T., and Wilken, P. (eds.). Oxford: Oxford University Press.
Inwood, B, and Gerson, L. P. (1994). *The Epicurus Reader*. Cambridge: Hackett Publishing.
Johnson, A., and Redish, A. D. (2007). Neural ensembles in CA3 transiently encode paths forward of the animal at a decision point. *Journal of Neuroscience* 27.45: 12176–89.
Jouvet, M. (1962). Recherches sur les structures nerveuses et les mécanismes responsables des différentes phases du sommeil physiologique. *Archives italiennes de biologie* 100: 125–206.
———. (1965a). Behavioral and EEG effects of paradoxical sleep deprivation in the cat. In *Proceedings of the 23rd International Congress of Physiological Sciences* (Vol. 4), Noble, D. (ed.). Excerpta Medica.
———. (1965b). Paradoxical sleep—a study of its nature and mechanisms. *Progress in Brain Research* 18: 20–62.
———. (1979). What does a cat dream about? *Trends in Neurosciences* 2: 280–82.
———. (2000). *The Paradox of Sleep: The Story of Dreaming*. Cambridge: MIT Press.

Kahan, T. L. (1994). Lucid dreaming as metacognition: Implications for cognitive science. *Consciousness and Cognition* 3: 246–64.

Kahane, G., and Savulescu, J. (2009). Brain damage and the moral significance of consciousness. *Journal of Medicine and Philosophy* 34: 6–26.

Karlsson, M. P., and Frank, L. M. (2009). Awake replay of remote experiences in the hippocampus. *Nature Neuroscience* 12: 913–18.

Karmanova, I. G. (1982). *Evolution of Sleep: Stages of the Formation of the "Wakefulness-sleep" Cycle in Vertebrates*. Basel: Karger.

Karmanova, I. G., and Lazarev, S. G. (1979). Stages of sleep evolution (facts and hypotheses). *Waking and Sleeping* 3: 137–47.

Kelly, D. (2018). The untold truth of Koko. *Grunge*. June 22, 2018. https://www.grunge.com/126879/the-untold-truth-of-koko/.

Kilian, A., Yaman, S., von Fersen, L., and Güntürkün, O. (2003). A bottlenose dolphin discriminates visual stimuli differing in numerosity. *Animal Learning & Behavior* 31: 133–42.

King, B. J. (2011). Are apes and elephants persons? In *Search of Self: Interdisciplinary Perspectives on Personhood*, Van Huyssteen, J. W., and Wiebe, E. P. (eds.), 70–82. Grand Rapids: Eerdmans Publishing.

Kingdom, S. (2017). The elephant orphans of Zambia. *Africa Geographic*. Accessed September 26, 2019. https://africageographic.com/blog/elephant-orphans-zambia/.

Kinnaman, A. J. (1902). Mental life of two Macacus rhesus monkeys in captivity. Part II. *American Journal of Psychology* 13: 173–218.

Kirmayer, L. J. (2009). Nightmares, neurophenomenology and the cultural logic of trauma. *Culture, Medicine, and Psychiatry* 33: 323–31.

Kittay, E. F. (2009). The personal is philosophical is political: A philosopher and mother of a cognitively disabled person sends notes from the battlefield. *Metaphilosophy* 40: 606–27.

Kittay, E. F., and Carlson, L., eds. (2010). *Cognitive Disability and Its Challenge to Moral Philosophy*. Hoboken: John Wiley & Sons.

Klein, C. (2007). An imperative theory of pain. *Journal of Philosophy* 104: 517–32.

Klein, M. (1963). Etude polygraphique et phylogénétique des différents états de sommeil. Thèse de Médecine. Lyon.

Knierim, J. J. (2009). Imagining the possibilities: ripples, routes, and reactivation. *Neuron* 63: 421–23.

Knobe, J., and Prinz, J. (2008). Intuitions about consciousness: Experimental studies. *Phenomenology and the Cognitive Sciences* 7: 67–83.

Kockelmans, J. J. (1994). *Edmund Husserl's phenomenology.* West Lafayette, Indiana: Purdue University Press.

Kornell, N., Son, L. K., and Terrace, H. S. (2007). Transfer of metacognitive skills and hint seeking in monkeys. *Psychological Science* 18.1: 64–71.

Korsgaard, C. (2018). *Fellow Creatures: Our Obligations to the Other Animals.* Oxford: Oxford University Press.

Kriegel, U. (2017). Dignity and the phenomenology of recognition-respect. In *Emotional Experience: Ethical and Social Significance,* Drummond, J. J., and Rinofner-Kreidl, S. (eds.), 121–36. Lanham, Maryland: Rowman & Littlefield.

———. (2019). The value of consciousness. *Analysis* 79: 503–20.

———. (n.d.). The value of consciousness: A propaedeutic. Accessed July 23, 2020. https://uriahkriegel.com/userfiles/downloads/propaedeutic.pdf.

Kunzendorf, R. G. (2016). *On the Evolution of Conscious Sensation, Conscious Imagination, and Consciousness of Self.* New York: Routledge.

LaBerge, S. (1985). *Lucid dreaming.* New York: Tarcher.

LaBerge, S., and DeGracia, D. J. (2000). Varieties of lucid dreaming experience. In *Individual Differences in Conscious Experience,* Kunzendorf, G., and Wallace, B. (eds.), 269–307. Amsterdam: John Benjamins Publishing Company.

Lacrampe, C. (2002). *Dormir, rêver: Le sommeil des animaux.* Paris: Iconoclaste.

LeDoux, J. E. (2013). The slippery slope of fear. *Trends in Cognitive Sciences* 17: 155–56.

Lee, A. (2019). Is consciousness intrinsically valuable? *Philosophical Studies* 176: 655–71.

Lesku, J. A., Meyer, L. C., Fuller, A., Maloney, S. K., Dell'Omo, G., Vyssotski, A. L., and Rattenborg, N. C. (2011). Ostriches sleep like platypuses. *PloS One* 6: e23203.

Leung, L. C., Wang, G. X., Madelaine, R., Skariah, G., Kawakami, K., Deisseroth, K., and Mourrain, P. (2019). Neural signatures of sleep in zebrafish. *Nature,* 571.7764: 198–204.

Levinas, E. (1979). *Totality and Infinity: An Essay on Exteriority.* New York: Springer.

———. (1981). *Otherwise than Being or beyond Essence.* New York: Springer.

Levy, N. (2014). The value of consciousness. *Journal of Consciousness Studies* 21: 127–38.

Levy, N., and Savulescu, J. (2009). Moral significance of phenomenal consciousness. *Progress in Brain Research,* 177: 361–70.

Lillard, A. S. (1994). Making sense of pretence. In *Children's early understanding of mind: Origins and development*, Lewis, C., and Mitchell, P. (eds.), 211–34. New York: Psychology Press.

Lindsay, W. L. (1879). *Mind in the Lower Animals in Health and Disease*. New York: Appleton.

Llinás, R. R., and Paré, D. (1991). Of dreaming and wakefulness. *Neuroscience* 44.3: 521–35.

Lohmar, D. (2007). How do primates think? Phenomenological analyses of non-language systems of representation in higher primates and humans. In *Phenomenology and the Non-human Animal*, Painter, C. and Lotz, C. (eds.), 57–74. New York: Springer.

———. (2016). *Denken ohne sprache: phänomenologie des nicht-sprachlichen denkens bei mensch und tier im licht der evolutionsforschung, primatologie und neurologie*. New York: Springer-Verlag.

Lopresti-Goodman, S. M., Kameka, M., and Dube, A. (2013). Stereotypical behaviors in chimpanzees rescued from the African bushmeat and pet trade. *Behavioral Sciences* 3.1: 1–20.

Louie, K., and Wilson, M. A. (2001). Temporally structured replay of awake hippocampal ensemble activity during rapid eye movement sleep. *Neuron* 29.1: 145–56.

Luce, G. (1966). Current research on sleep and dreams. Public Health Service Publication No. 1389. National Institute of Mental Health.

Lucretius, C. T. (1910). *On the Nature of Things*. Bailey, C. (trans.). Oxford: Oxford University Press.

Lyamin, O. I., Shpak, O. V., Nazarenko, E. A., and Mukhametov, L. M. (2002). Muscle jerks during behavioral sleep in a beluga whale (*Delphinapterus leucas L.*). *Physiology & Behavior* 76.2: 265–70.

Lyn, H., Greenfield, P., and Savage-Rumbaugh, S. (2006). The development of representational play in chimpanzees and bonobos: Evolutionary implications, pretense, and the role of interspecies communication. *Cognitive Development* 21.3: 199–213.

Malcolm, N. (1956). Dreaming and skepticism. *Philosophical Review* 65: 14–37.

———. (1959). *Dreaming*. New York: Routledge.

Malinowski, J. E., Scheel, D., and McCloskey, M. (2021). Do animals dream? *Consciousness and Cognition* 95: 103214.

Mallatt, J., and Feinberg, T. E. (2016). Insect consciousness: Fine-tuning the hypothesis. *Animal Sentience* 1.9: 10.

Manger, P. R., and Siegel, J. M. (2020). Do all mammals dream? *Journal of Comparative Neurology* 528: 1–39.

Mann, J. (2018). *Deep Thinkers: Inside the Minds of Whales, Dolphins, and Porpoises*. Chicago: University of Chicago Press.

Mann, W. E. (1983). Dreams of immorality. *Philosophy* 58: 378–85.

Masson, J. M. (2009). *When Elephants Weep: The Emotional Lives of Animals*. New York: Delta.

Matsuzawa, T. (2009). Symbolic representation of number in chimpanzees. *Current Opinion in Neurobiology* 19.1: 92–98.

———. (2011). Log doll: Pretense in wild chimpanzees. In *The Chimpanzees of Bossou and Nimba*. Matsuzawa, T., Humle, T., and Sugiyama, T. (eds.), 131–35. New York: Springer.

Matthews, G. B. (1981). On being immoral in a dream. *Philosophy* 56: 47–54.

Merleau-Ponty, M. (2013). *Phenomenology of Perception*. New York: Routledge.

Metzinger, T. (2003). *Being No One: The Self-model Theory of Subjectivity*. Cambridge: MIT Press.

———. (2009). *The Ego Tunnel: The Science of the Mind and the Myth of the Self*. New York: Basic Books.

Miller, G. A. (1962). *Psychology: The Science of Mental Life*. London: Pelican Books.

Mills, W. (1889). *A Textbook of Animal Physiology: With introductory chapters on general biology and a full treatment of reproduction, for students of human and comparative (veterinary) medicine and of general biology*. Boston: D. Appleton and Company.

Mitchell, R. W. (2016). Can animals imagine? In *Routledge Handbook of Philosophy of Imagination*, Kind, A. (ed.), 326–38. New York: Routledge.

Montaigne, M. (1877). *The Complete Essays of Michael de Montaigne*. Cotton, C. (trans.), Hazlitt, W. (ed.). https://gutenberg.org/files/3600/3600-h/3600-h.htm.

Montangero, J. (2012). Dream thought should be compared with waking world simulations: A comment on Hobson and colleagues' paper on dream logic. *Dreaming* 22: 70–73.

Morin, R. (2015). A conversation with Koko the gorilla: An afternoon spent with the famous gorilla who knows sign language and the scientist who taught her how to talk. *Atlantic*. August 28, 2015. https://www.theatlantic.com/technology/archive/2015/08/koko-the-talking-gorilla-sign-language-francine-patterson/402307/.

Morse, D. D., and Danahay, M. A., eds. (2017). *Victorian Animal Dreams: Representations of Animals in Victorian Literature and Culture*. New York: Routledge.

Mukobi [previously Williams], K. (1995). Comprehensive Nighttime Activity Budgets of Captive Chimpanzees (*pan troglodytes*). Master's thesis, Central Washington University.

Nagel, T. (1974). What is it like to be a bat? *Philosophical Review* 83: 435–50.

Newen, A., and Bartels, A. (2007). Animal minds and the possession of concepts. *Philosophical Psychology* 20.3: 283–308.

Nicol, S. C., Andersen, N. A., Phillips, N. H., & Berger, R. J. (2000). The echidna manifests typical characteristics of rapid eye movement sleep. *Neuroscience Letters* 283.1: 49–52.

Noë, A. (2009). *Out of Our Heads: Why You Are Not Your Brain, and Other Lessons from the Biology of Consciousness*. New York: Macmillan.

O'Neill, J., Senior, T. J., Allen, K., Huxter, J. R., and Csicsvari, J. (2008). Reactivation of experience-dependent cell assembly patterns in the hippocampus. *Nature Neuroscience* 11.2: 209–15.

O'Neill, J., Senior, T., and Csicsvari, J. (2006). Place-selective firing of CA1 pyramidal cells during sharp wave/ripple network patterns in exploratory behavior. *Neuron* 49.1: 143–55.

O'Neill, O. (1997). Environmental values, anthropocentrism and speciesism. *Environmental Values* 6.2: 127–42.

Occhionero, M., and Cicogna, P. (2016). Phenomenal consciousness in dreams and in mind wandering. *Philosophical Psychology* 29.7: 958–66.

Ólafsdóttir, H. F., Barry, C., Saleem, A. B., Hassabis, D., and Spiers, H. J. (2015). Hippocampal place cells construct reward related sequences through unexplored space. *Elife* 4: e06063.

Ólafsdóttir, H. F., Bush, D., and Barry, C. (2018). The role of hippocampal replay in memory and planning. *Current Biology* 28.1: R37–R50.

Olthof, A., and Roberts, W. A. (2000). Summation of symbols by pigeons (*Columba livia*): The importance of number and mass of reward items. *Journal of Comparative Psychology* 114.2: 158.

Pagel, J. F., and Kirshtein, P. (2017). *Machine Dreaming and Consciousness*. Cambridge: Academic Press.

Pantani, M., Tagini, T., and Raffone, A. (2018). Phenomenal consciousness, access consciousness and self across waking and dreaming: bridging phenomenology and neuroscience. *Phenomenology and the Cognitive Sciences* 17.1: 175–97.

Pastalkova, E., Itskov, V., Amarasingham, A., and Buzsáki, G. (2008). Internally generated cell assembly sequences in the rat hippocampus. *Science* 321.5894: 1322–27.

Pearson, K. A., and Large, D. (2006). *The Nietzsche Reader*. Hoboken: Blackwell.

Peña-Guzmán, D. M. (2017). Can nonhuman animals commit suicide? *Animal Sentience* 20.1: 1–24.

———. (2018). Can nondolphins commit suicide? *Animal Sentience* 20.20: 1–22.

Pepperberg, I. M. (2012). Further evidence for addition and numerical competence by a Grey parrot (*Psittacus erithacus*). *Animal Cognition* 15.4: 711–17.

———. (2013). Abstract concepts: Data from a grey parrot. *Behavioural Processes* 93: 82–90.

Plato. (2000). *The Republic*. Ferrari, G. (ed.). Cambridge: Cambridge University Press.

Poovey, M. (1998). *A History of the Modern Fact: Problems of Knowledge in the Sciences of Wealth and Society*. Chicago: University of Chicago Press.

Preston, E. (2019). Was Heidi the octopus really dreaming? *New York Times*, October 8, 2019.

Ramsey, J. K., and McGrew, W. C. (2005). Object play in great apes. In *The Nature of Play: Great Apes and Humans*. Pellegrini, A. D., and Smith P. K. (eds.), 89–112. New York: Guilford Press.

Raymond, E. L. (1990). *An Examination of Private Signing in Deaf Children in a Naturalistic Environment*. Doctoral dissertation, Central Washington University.

Regan, T. (2004). *The Case for Animal Rights*. Berkeley: University of California Press.

Rescorla, M. (2009). Chrysippus' dog as a case study in non-linguistic cognition. In *The Philosophy of Animal Minds*, Lurz, R. (ed.), 52–71. Cambridge: Cambridge University Press.

Revonsuo, A. (2000). The reinterpretation of dreams: An evolutionary hypothesis of the function of dreaming. *Behavioral and Brain Sciences* 23: 877–901.

———. (2005). The self in dreams. In *The Lost Self: Pathologies of the Brain and Identity*, Feinberg, T. and Keenan, J. P. (eds.), 206–19. Oxford: Oxford University Press.

———. (2006). *Inner Presence: Consciousness as a Biological Phenomenon*. Cambridge: MIT Press.

Ridley, Matt. (2003). *Nature via Nurture: Genes, Experience, and What Makes Us Human*. New York: Harper Collins.

Ridley, R. M., Baker, H. F., Owen, F., Cross, A. J., and Crow, T. J. (1982). Behavioural and biochemical effects of chronic amphetamine treatment in the vervet monkey. *Psychopharmacology* 78.3: 245–51.

Robbins, T. W. (2017). Animal models of hallucinations observed through the modern lens. *Schizophrenia Bulletin* 43.1: 24–26.

Rock, A. (2004). *The Mind at Night: The New Science of How and Why We Dream*. New York: Basic Books.

Romanes, G. (1883). *Mental Evolution in Animals*. London: Kegan Paul Trench & Co.

Rosenthal, D. (1997). A theory of consciousness. In *The Nature of Consciousness*, Block, N. and Flanagan, O. J. (eds.), 729–54. Cambridge: MIT Press.

———. (2005). *Consciousness and Mind*, Cambridge: Clarendon Press.

Rotenberg, V. S. (1993). REM sleep and dreams as mechanisms of the recovery of search activity. In *The Functions of Dreaming*, Moffitt, A., Kramer, M., and Hoffmann, R. (eds.), 261–92. Albany: SUNY Press.

Rowe, K., Moreno, R., Lau, T. R., Wallooppillai, U., Nearing, B. D., Kocsis, B., Quattrochi, J., Hobson, J. A., and Verrier, R. L. (1999). Heart rate surges during REM sleep are associated with theta rhythm and PGO activity in cats. *American Journal of Physiology-Regulatory, Integrative and Comparative Physiology* 277.3: R843-R849.

Rowlands, M. (2009). *Animal Rights: Moral Theory and Practice*. London: Palgrave.

San Martín, J., and Peñaranda, M.L.P. (2001). Animal life and phenomenology. In *The Reach of Reflection: Issues for Phenomenology's Second Century*, Vol. 2, Crowell, S., Embree, L., and Julia, S. J. (eds.), 342–63. Boca Raton, Florida: Florida Atlantic University, the Center for Advanced Research in Phenomenology.

Santayana, G. (1940). *The Philosophy of George Santayana*, Volume 2, Schilpp, P. A. (ed.). New York: Tudor Publishing Company.

Sartre, J. P. (2004). *The Imaginary: A Phenomenological Psychology of the Imagination*. Hove: Psychology Press.

Schmitt, V., and Fischer, J. (2009). Inferential reasoning and modality dependent discrimination learning in olive baboons (*Papio hamadryas anubis*). *Journal of Comparative Psychology* 123: 316.

Searle, J. R. (1998). How to study consciousness scientifically. *Philosophical Transactions of the Royal Society of London. Series B: Biological Sciences* 353: 1935–42.

Sebastián, M. Á. (2014a). Dreams: An empirical way to settle the discussion between cognitive and non-cognitive theories of consciousness. *Synthese* 2: 263–85.

———. (2014b). Not a HOT dream. In *Consciousness Inside and Out: Phenomenology, Neuroscience, and the Nature of Experience*. Brown, R. (ed.), 415–32. New York: Springer.

Shafton, A. (1995). *Dream Reader: Contemporary Approaches to the Understanding of Dreams*. Albany: SUNY Press.

Shepherd, Joshua. (2018). *Consciousness and Moral Status*. Oxfordshire: Taylor & Francis.

Shurley, J. T., Serafetinides, E. A., Brooks, R. E., Elsner, R., Kenney, D. W. (1969). Sleep in Cetaceans: I. The pilot whale, *Globicephala scammony*. *Psychophysiology* 6: 230.

Siebert, C. (2011). Orphans no more. *National Geographic* 220.3: 40–65.

Siegel, J. M., Manger, P. R., Nienhuis, R., Fahringer, H. M., and Pettigrew, J. D. (1998). Monotremes and the evolution of rapid eye movement sleep. *Philosophical Transactions of the Royal Society of London. Series B: Biological Sciences* 353.1372: 1147–57.

Siegel, J. M., Manger, P. R., Nienhuis, R., Fahringer, H. M., Shalita, T., and Pettigrew, J. D. (1999). Sleep in the platypus. *Neuroscience* 91.1: 391–400.

Siegel, R. K. (1973). An ethologic search for self-administration of hallucinogens. *International Journal of the Addictions* 8.2: 373–93.

Siegel, R. K., Brewster, J. M., and Jarvik, M. E. (1974). An observational study of hallucinogen-induced behavior in unrestrained *Macaca mulatta*. *Psychopharmacologia* 40.3: 211–23.

Siegel, R. K., and Jarvik, M. E. (1975). Drug-induced hallucinations in animals and man. In *Hallucinations: Behavior, Experience and Theory*, Siegel R. K. and West, L. J. (eds.), 163–95. Hoboken: John Wiley & Sons.

Siewert, C. (1994). Speaking up for consciousness. In *Current Controversies in Philosophy of Mind*, Kriegel, U. (ed.), 199–221. New York: Routledge.

———. (1998). *The Significance of Consciousness*. Princeton: Princeton University Press.

Simondon, G. (2011). *Two Lessons on Animal and Man*. Minnesota: University of Minnesota Press.

Singer, Peter. (1995). *Animal Liberation*. New York: Random House.

Smith, J. D. (2009). The study of animal metacognition. *Trends in Cognitive Sciences* 13.9: 389–96.

Smith, J. D., and Washburn, D. A. (2005). Uncertainty monitoring and metacognition by animals. *Current Directions in Psychological Science* 14: 19–24.

Smith, J. D., Couchman, J. J., and Beran, M. J. (2012). The highs and lows of theoretical interpretation in animal-metacognition research. *Philosophical Transactions of the Royal Society B: Biological Sciences* 367: 1297–1309.

Solms, M. (2021). *The Hidden Spring: A Journey to the Source of Consciousness*. New York: WW Norton & Company.

Sosa, E. (2005). Dreams and philosophy. *Proceedings and Addresses of the American Philosophical Association* 79.2: 7–18.

Sparrow, R. (2010). A not-so-new eugenics: Harris and Savulescu on human enhancement. *Asian Bioethics Review* 2.4: 288–307.

Stahel, C. D., Megirian, D., and Nicol, S. C. (1984). Sleep and metabolic rate in the little penguin, *Eudyptula minor*. *Journal of Comparative Physiology B* 154.5: 487–94.

Starr, Michelle. (2019). Watch the Mesmerising Colour Shifts of a Sleeping Octopus. Online Video. Science Alert, September 27, 2019. https://www.sciencealert.com/watch-the-mesmerising-colour-shifts-of-a-sleeping-octopus.

Stein, E. (1989). *Zum problem der einfühlung* [*On the problem of empathy*]. Stein, W. (trans.). Washington, DC: ICS Publications.

Steiner, G. (1983). The Historicity of Dreams (Two questions to Freud). *Salmagundi* 61: 6–21.

Stephan, A. (1999). Are animals capable of concepts? *Erkenntnis* 51.1: 583–96.

Stone, C. D. (2010). *Should Trees Have Standing?: Law, Morality, and the Environment*. Oxford: Oxford University Press.

Strawson, G. (2009). *Mental Reality, with a New Appendix*. Cambridge: MIT Press.

Tayler, C. K., and Saayman, G. S. (1973). Imitative behaviour by Indian Ocean bottlenose dolphins (*Tursiops aduncus*) in captivity. *Behaviour* 44: 286–98.

Thomas, N. J. (2014). The multidimensional spectrum of imagination: Images, dreams, hallucinations, and active, imaginative perception. *Humanities* 3.2: 132–84.

Thompson, E. (2007). *Mind in Life: Biology, Phenomenology, and the Sciences of Mind*. Cambridge: Harvard University Press.

———. (2015). *Waking, Dreaming, Being: Self and Consciousness in Neuroscience, Meditation, and Philosophy*. New York: Columbia University Press.

Uller, C., and Lewis, J. (2009). Horses (*Equus caballus*) select the greater of two quantities in small numerical contrasts. *Animal Cognition* 12.5: 733–38.

Underwood, E. (2016). Do sleeping dragons dream? *Science Magazine*. April 28, 2016. https://www.sciencemag.org/news/2016/04/do-sleeping-dragons-dream.

Uexküll, J. (2013). *A Foray into the Worlds of Animals and Humans: With a Theory of Meaning*. Minnesota: University of Minnesota Press.

Valatx, J. L., Jouvet, D., and Jouvet, M. (1964). EEG evolution of the different states of sleep in the kitten. *Electroencephalography and Clinical Neurophysiology* 17.3: 218–33.

Van Cantfort, T. E., Gardner, B. T., and Gardner, R. A. (1989). *Teaching Sign Language to Chimpanzees*. Albany: SUNY Press.

Van der Kolk, B. (2015). *The Body Keeps the Score: Brain, Mind, and Body in the Healing of Trauma*. London: Penguin.

Van Twyver, H., and Allison, T. (1972). A polygraphic and behavioral study of sleep in the pigeon (*Columba livia*). *Experimental Neurology* 35.1: 138–53.

Vanderheyden, W. M., George, S. A., Urpa, L., Kehoe, M., Liberzon, I., and Poe, G. R. (2015). Sleep alterations following exposure to stress predict fear-associated memory impairments in a rodent model of PTSD. *Experimental Brain Research* 233.8: 2335–46.

Varela, F. J. (1999). The specious present: A neurophenomenology of time consciousness. In *Naturalizing Phenomenology*, Petitot, J., Varela, F. J., Pachoud, B., & Roy, J.-M. (eds.), 266–314. Palo Alto: Stanford University Press.

Visanji, N. P., Gomez-Ramirez, J., Johnston, T. H., Pires, D., Voon, V., Brotchie, J. M., and Fox, S. H. (2006). Pharmacological characterization of psychosis-like behavior in the MPTP-lesioned nonhuman primate model of Parkinson's disease. *Movement Disorders: Official Journal of the Movement Disorder Society* 21.11: 1879–91.

Voltaire. (1824). Imagination. In *A Philosophical Dictionary*, Hunt, J., & Hunt, H. L. (eds.), 116–24. New York: Alfred A. Knopf.

Vonk, J., and Beran, M. J. (2012). Bears "count" too: Quantity estimation and comparison in black bears, *Ursus americanus*. *Animal Behaviour* 84.1: 231–38.

Voss, U. (2010). Lucid dreaming: Reflections on the role of introspection. *International Journal of Dream Research* 3.1: 52–53.

Voss, U., and Hobson, A. (2014). What is the state-of-the-art on lucid dreaming? Recent advances and questions for future research. In *Open MIND*, Metzinger, T. &Windt, J. M. (eds.), 38(T). Frankfurt: MIND Group.

Walker, J. M., and Berger, R. J. (1972). Sleep in the domestic pigeon (*Columba livia*). *Behavioral Biology* 7.2: 195–203.

Walsh, R. N., and Vaughan, F. (1992). Lucid dreaming: Some transpersonal implications. *Journal of Transpersonal Psychology* 24: 19.

Walton, K. L. (1990). *Mimesis as Make-believe: On the Foundations of the Representational Arts*. Cambridge: Harvard University Press.

Warren, M. A. (1997). *Moral Status: Obligations to Persons and Other Living Things*. Cambridge: Clarendon Press.

Watanabe, S., and Huber, L. (2006). Animal logics: Decisions in the absence of human language. *Animal Cognition* 9.4: 235–45.

West, R. E., and Young, R. J. (2002). Do domestic dogs show any evidence of being able to count? *Animal Cognition* 5.3: 183–86.

Willett, C. (2014). *Interspecies Ethics*. New York: Columbia University Press.

Windt, J. M. (2010). The immersive spatiotemporal hallucination model of dreaming. *Phenomenology and the Cognitive Sciences* 9: 295–316.

Windt, J. M. (2015). *Dreaming: A Conceptual Framework for Philosophy of Mind and Empirical Research*. Cambridge: MIT Press.

Windt, J. M., and Metzinger, T. (2007). The philosophy of dreaming and self-consciousness: What happens to the experiential subject during the dream state. In *The New Science of Dreaming, Volume 3: Cultural and Theoretical Perspectives*, Barrett, D., and McNamara, P. (eds.), 193–247. Westport: Praeger Publishers.

Windt, J. M., and Voss, U. (2018). Spontaneous thought, insight, and control in lucid dreams. In *The Oxford Handbook of Spontaneous Thought: Mind-Wandering, Creativity, and Dreaming*, Fox, K., and Christoff, K. (eds), 385–410. Oxford: Oxford University Press.

Wittgenstein, L. (1958). *Philosophical Investigations*. Anscombe, GEM (trans.). Oxford: Oxford University Press.

Wolfe, C. (2013). Learning from Temple Grandin, or, animal studies, disability studies, and who comes after the subject. In *Re-Imagining Nature, Environmental Humanities and Ecosemiotics*, Carey, J., Cohen, J. J., Faull, K. M., Maran, T., Moran, D., Oleksa, M., Radding, C., Reese, S., Shanley, K. W., and Wolfe, C. (eds.), 91–107. Lewisburg: Bucknell University Press.

Yu, B., Cui, S. Y., Zhang, X. Q., Cui, X. Y., Li, S. J., Sheng, Z. F., Cao, Q., Huang, Y. L., Xu, Y. P., Lin, Z. G., and Yang, G. (2015). Different neural circuitry is involved in physiological and psychological stress-induced PTSD-like "nightmares" in rats. *Scientific Reports* 5.1: 1–14.

———. (2016). Mechanisms underlying footshock and psychological stress-induced abrupt awakening from posttraumatic "nightmares." *International Journal of Neuropsychopharmacology* 19: 1–6.

Zahavi, D. (2014). *Self and Other: Exploring Subjectivity, Empathy, and Shame*. Oxford: Oxford University Press.

Zepelin, H. (1994). Mammalian Sleep. In *Principles and Practice of Sleep Medicine*, Kryger, M. H., Roth, T., and Dement, W. C. (eds.), 69–80. Philadelphia: W.B. Saunders Company.

Zhang, Q. (2009). A computational account of dreaming: Learning and memory consolidation. *Cognitive Systems Research* 10.2: 91–101.

INDEX